Ramsey Theory for Discrete Structures

Hans Jürgen Prömel

Ramsey Theory for Discrete Structures

 Springer

Hans Jürgen Prömel
Technische Universität Darmstadt
Darmstadt, Germany

ISBN 978-3-319-34486-7   ISBN 978-3-319-01315-2 (eBook)
DOI 10.1007/978-3-319-01315-2
Springer Cham Heidelberg New York Dordrecht London

Mathematics Subject Classification (2010): 05-02

Printed on acid-free paper

Springer is part of Springer Science+Business Media (www.springer.com)

Foreword

In spring 1986, I started as a PhD student at the Institute of Operations Research at the University of Bonn. At that time, Hans Jürgen Prömel was a PostDoc there and quickly became my mentor. Naturally, I was interested in his research and thus involved myself in proofreading large parts of his Habilitation thesis on *Ramsey Theory for Discrete Structures*. I still vividly remember that time: I had never been in touch with Ramsey theory before (and in fact never after), but I thoroughly enjoyed that exposure to a new area. The thesis was meant as a first part to a forthcoming monograph on Ramsey theory to be written jointly with Bernd Voigt. And with this in mind, the thesis was written. It was much more than a collection of research papers. It was an introduction to Ramsey Theory in which the author conveyed both his love for the field and his tremendous insights: a combination that was extremely fascinating for a first year PhD student.

Unfortunately, as it so happens, life had new challenges for both authors before the planned book was written. Bernd Voigt left academia and started a new and successful career in industry. Hans Jürgen stayed in an academic environment, but moved his focus, first just within research and then more and more towards management, culminating in his election as president of TU Darmstadt in 2007.

In September 2013, Hans Jürgen will celebrate his 60th birthday. We will host a colloquium in his honor at ETH Zürich. While thinking about a "birthday present," memories of my first year as a PhD student came back, and the idea evolved that a valuable gift to him, as well as the community, could be to finish his book, not as the grand monograph as it was once planned but never finished but as a thorough introduction to Ramsey theory as provided by his Habilitation thesis that I once read and loved.

Luckily, Springer was immediately very supportive of this idea. What you have in your hands is what came out of this project: a second edition of a book whose first edition was never published. Parts I–III are essentially the same as they were in the thesis. Some references and paragraphs were added, a few sections removed, and some of the proofs enhanced with more details wherever that seemed appropriate. Parts IV and V of the thesis are now merged. Some of the more technical and lengthy deterministic constructions were replaced by a section on random Ramsey theory.

Finally, a chapter on the recent developments of the polymath project was added, i.e., a chapter with a combinatorial proof of the density Hales-Jewett theorem.

This project would not have been possible without the tremendous help and assistance from my student Rajko Nenadov. He not only spotted typos and criticized the text, but more often than not, he came up with new ideas and suggestions of how to improve the presentation of the proofs and chapters at hand. In particular, Part IV would look much different without his invaluable help. In addition to Rajko several other members of my group helped me finish this project. I am deeply grateful to all of them. A special thanks to Andreas Noever whose careful proofreading of the final version ensured that there are now much fewer typos than there were before. Last but not least, a big thank you to Springer for making this project possible!

Zürich, Switzerland   Angelika Steger
June 2013

Preface

Man kann mit der Frage beginnen: Was ist Abstraktion, und welche Rolle spielt sie im begrifflichen Denken? Als Antwort kann man etwa formulieren: Abstraktion bezeichnet die Möglichkeit, einen Gegenstand oder eine Gruppe von Gegenstanden unter einem Gesichtspunkt unter Absehen von allen anderen Gegenstandseigenschaften zu betrachten.

(Heisenberg 1960)

In the 1890s, D. Hilbert, in connection with investigations on the irreducibility of rational functions, proved a Ramsey-type result nowadays known as Hilbert's cube lemma. Some 25 years later, I. Schur showed in reproving a theorem of Dickson on a modular version of Fermat's conjecture that if the positive integers are finitely colored, one color class contains x, y, and z with $x+y = z$. A conjecture concerning the distribution of quadratic residues, respectively nonresidues modulo p, led Schur to a question on arithmetic progressions. This problem was solved in 1927 by B. L. van der Waerden and the corresponding theorem became famous as van der Waerden's theorem on arithmetic progressions.

Around the same time the English mathematician F. P. Ramsey tried to give a decision procedure for propositional logic. The need for such procedures, we would say algorithms in the present-day terminology, arose with the crisis of the foundations of mathematics around 1900. It was more or less the theory of sets and the arithmetization of analysis which led to this crisis. In response, the programs of Russell and Whitehead, of Hilbert, and of Brouwer called for a new foundation trying to overwhelm the doubtful principles of mathematics of that time.

It is a kind of irony that a purely mathematical result from Ramsey's paper, an astonishing generalization of the pigeonhole principle, has proved to be of so much greater consequence than the metamathematical investigations for which they were made as tools. Even more, for this result Ramsey became eponymous for a part of discrete mathematics known as Ramsey theory.

These four roots of Ramsey theory were established for different reasons, unaware of the other. A first culmination point, then, was obtained with the work of R. Rado, a doctoral student of Schur. In his *Studien zur Kombinatorik* and several subsequent papers, Rado unified and extended the results of Hilbert, Schur, and van der Waerden giving a complete characterization of those systems of linear equations which are partition regular.

Quite independently from this direction of research, there was a profound development based on Ramsey's theorem, which is closely connected with the name of P. Erdős. A kind of first step in popularizing Ramsey's theorem was an application to a combinatorial problem in geometry due to Erdős and Szekeres: "I am sure that this paper had a strong influence on both of us. Paul with his deep insight recognized the possibilities of a vast unexplored territory and opened up a new world of combinatorial set theory and combinatorial geometry" (Szekeres 1973). But it took until the middle of the 1960s when A. W. Hales and R. I. Jewett revealed the combinatorial core of van der Waerden's theorem on arithmetic progressions proving a kind of pigeonhole principle for parameter sets. Some years later, R. L. Graham and B. L. Rothschild extended Hales-Jewett's result in a remarkable way. They established a complete analogue to Ramsey's theorem for the structure of parameter sets and, as it turns out, Ramsey's theorem itself is an immediate consequence of the Graham-Rothschild theorem. But the concept of parameter sets does not only glue together arithmetic progressions and finite sets. It also provides a natural framework for seemingly different structures like Boolean lattices, partition lattices, hypergraphs and Deuber's (m, p, c)-sets, just to mention a few. So, to a certain extent the Graham-Rothschild theorem can be viewed as a starting point of Ramsey theory for discrete structures:

> Dies kann also beim Vorgang der Abstraktion geschehen: Der im Prozeß der Abstraktion gebildete Begriff gewinnt ein eigenes Leben, er läßt eine unerwartete Fülle von Formen oder ordnenden Strukturen aus sich entstehen, die sich später auch beim Verständnis der uns umgebenden Erscheinungen in irgendeiner Weise bewähren können (Heisenberg 1960).

The present work is organized as follows. In the first part, we give a more detailed discussion of the roots of Ramsey theory. Thereafter, we focus on three discrete structures: sets, parameter sets, and graphs.

The second part of this work contains a thorough discussion of the role of parameter sets in Ramsey theory. Originally, the idea was to find a combinatorial abstraction of linear and affine spaces over finite fields. This was motivated by a conjecture of G. C. Rota proposing a geometric analogue to Ramsey's theorem. But the impact of parameter sets goes far beyond the proof of Rota's conjecture. In Chap. 3 we present some definitions and several examples of structures which can be interpreted in terms of parameter sets. Chapters 4 and 5, then, contain the most fundamental Ramsey-type results for parameter sets, viz., Hales-Jewett's theorem and Graham-Rothschild's Ramsey theorem for n-parameter sets, as well as several applications thereof. Finally, in Chap. 6 we build upon the Graham-Rothschild theorem to obtain canonical versions of the aforementioned results.

In the third part, we go back to the most basic structure, to sets, and discuss developments which originate in Ramsey's theorem itself. One of the oldest areas in Ramseyean research is the study of Ramsey numbers which essentially starts with the paper of Erdős and Szekeres. We devote Chap. 7 to review old results as well as recent progress on Ramsey numbers and on the asymptotic behavior of the classical Ramsey functions. In Chap. 8 unprovability results are discussed. As it turns out, a slight variation of the finite Ramsey theorem is one of the first mathematical interesting examples for Gödel's incompleteness theorem. Chapter 9 presents product versions of Ramsey's theorem, whereas Chap. 10 covers a result on the necessity of irregularities of set partitions. In the final chapter of this part, we discuss extensions of Ramsey's theorem to larger cardinals, based on the profound work of Erdős, Hajnal, and Rado in this area.

Graphs and hypergraphs seem to be one of the most alive and exciting areas of research in Ramsey theory nowadays. In Chaps. 12 and 14, we present a complete solution of the Ramsey problem for finite graphs, respectively hypergraphs, closely connected with the names of Deuber, Nešetřil, and Rödl. Moreover, we introduce and develop an amalgamation technique for graphs and hypergraphs which is an essential tool in proving sparse and restricted Ramsey theorems. In between, in Chap. 13, we collect some results which are known for infinite graphs, mainly due to Erdős, Hajnal, and Pósa. In Chap. 14, we start to consider graphs and hypergraphs in a broader perspective. Ramsey's theorem for finite hypergraphs can be viewed as an induced version of Ramsey's theorem. Apparently Spencer was the first to consider graphs and hypergraphs which are defined on more complex structures than just sets, proving an induced version of van der Waerden's theorem. In this last part of Chap. 14 we introduce hypergraphs defined on parameter sets and prove an induced Graham-Rothschild theorem.

Sparse Ramsey theorems for graphs originate in investigations of graphs having large chromatic number and high girth. A complete solution to the problem was first given by Erdős using probabilistic means and later by Lovász via an explicit construction. In Chap. 16 we give an account on the probabilistic method for constructing more general sparse Ramsey configurations.

Several areas of Ramsey theory remain uncovered throughout this work, e.g., Euclidean Ramsey theory or topological Ramsey theory. We refer the interested reader to the excellent monograph *Ramsey Theory* of Graham et al. (1980), as well as to the forthcoming volume *Mathematics of Ramsey Theory* edited by Nešetřil and Rödl (1990). We also do not discuss any of the recent applications of Ramsey theory to computer science. Here we refer the reader, for example, to Alon and Maass (1986), Moran et al. (1985), Nešetřil (1984), and Pudlák (1990), just to mention a few.

Kehren wir zu der am Anfang gestellten Frage zurück. Der Zug zur Abstraktion in der Naturwissenschaft beruht also letzten Endes auf der Notwendigkeit, weiterzufragen, auf dem Streben nach einem einheitlichen Verständnis. ... die Menschen, die über die Natur nachdenken, fragen weiter, weil sie die Welt als Einheit begreifen, ihren einheitlichen Bau verstehen wollen. Sie bilden zu diesem Zweck immer umfassendere Begriffe deren Zusammenhang mit dem unmittelbaren sinnlichen Erlebnis nur schwer zu erkennen ist wobei aber das Bestehen eines solchen Zusammenhangs unabdingbare Voraussetzung dafür ist, daß die Abstraktion überhaupt noch Verständnis der Welt vermittelt. (Heisenberg 1960)

Bonn, Germany   Hans Jürgen Prömel
June 1987

Conventions

Definitions and basic terminology will be introduced throughout this work as needed. Here we will only agree on some general conventions to get started. Unless otherwise specified, numbers are nonnegative integers. In particular, $i \leq k$ is a shorthand notation for $0 \leq i \leq k$ and $[i, k]$ abbreviates the set $i, i+1, \ldots, k-1, k$ of integers.

Nonnegative integers are identified with the set of their predecessors, e.g., $k = \{0, \ldots, k-1\}$ which is the ordinal notation. The smallest infinite ordinal is denoted by ω, the set of nonnegative integers.

For X being a set, we denote by $[X]^k$ the set of all k-subsets of X, i.e.,

$$[X]^k = \{Y \subseteq X \mid |Y| = k\},$$

where $|Y|$ denotes the cardinality of Y. In particular, $[\omega]^k$ is the set of all k-element sets of nonnegative integers, and $[\omega]^\omega$ is the set of all infinite sets of nonnegative integers. The set of n-tuples, or words of length n, over an alphabet A is denoted by A^n.

If $\Delta : n \to r$ is a mapping and $M \subseteq n$ a subset of $n = \{0, \ldots, n-1\}$, then $\Delta{\upharpoonright}M$ denotes the restriction of Δ to M. More precisely, the mapping $\Delta{\upharpoonright}M$ is defined by $(\Delta{\upharpoonright}M)(i) = \Delta(i)$ for every $i \in M$. In Ramsey theory, such mappings are usually called colorings, and if their range is $r = \{0, \ldots, r-1\}$, they are r-colorings.

A basic principle in Ramsey theory is the pigeonhole principle (Dirichlets Schubfachprinzip):

(*Finite version*) Let m and r be positive integers and $n \geq r \cdot (m-1) + 1$. Then for every r-coloring $\Delta : n \to r$, there exists an m-subset $M \in [n]^m$ such that $\Delta{\upharpoonright}M$ is a constant coloring.

(*Infinite version*) Let r be a positive integer. Then for every r-coloring $\Delta : \omega \to r$, there exists an infinite subset F in $[\omega]^\omega$ such that $\Delta{\upharpoonright}F$ is a constant coloring.

To a certain extent, all results contained in this work can be viewed as generalizations of the pigeonhole principle.

We denote by $\mathbb{N}$ the set of positive integers. The integers, the rationals, and reals will be denoted by $\mathbb{Z}$, $\mathbb{Q}$, and $\mathbb{R}$, respectively. The greatest integer not greater than the real number x will be written as $\lfloor x \rfloor$ and the least integer not less than x as $\lceil x \rceil$. We use Landau's notation $O(f(n))$ for a term which, when divided by $f(n)$, remains bounded as $n \to \infty$. Similarly, $h(n) = o(g(n))$ means that $h(n)/g(n) \to 0$ as $n \to \infty$, and, for convenience, $h(n) = \Omega(g(n))$ abbreviates that $g(n)/h(n)$ is bounded as $n \to \infty$. Finally, $\log x$ denotes the binary logarithm of x, whereas $\ln x$ denotes the natural logarithm of x, i.e., the logarithm to the base e.

Contents

Part III   Back to the Roots: Sets

Part I
Roots of Ramsey Theory

Chapter 1
Ramsey's Theorem

1.1 Frank Plumpton Ramsey

Frank Plumpton Ramsey was an extraordinary man. He was born in 1903 in Cambridge as the elder son of A.S. Ramsey who was a mathematician and President of Magdalene College. His younger brother Michael went on to become Archbishop of Canterbury.

While, and even before, studying mathematics at Trinity College, Ramsey was deeply influenced by some of the brilliant Cambridge thinkers of that time: Ramsey's early work in logic and philosophy of mathematics was strongly influenced by Bertrand Russell and Ludwig Wittgenstein. His *Foundations of mathematics* (Ramsey 1926a) is "an attempt to reconstruct the system of" (Russell and Whitehead's) "Principia Mathematica so that its blemishes may be avoided but its excellencies retained", his *Mathematical logic* (Ramsey 1926b) is a defense of logicism "against formalism of Hilbert and the intuitionism of Brouwer" (quoted from Braithwaite 1931). However, "neither Whitehead and Russell nor Ramsey succeeded in attaining the logicistic goal constructively" (quoted from Kleene 1952). Ramsey translated Wittgenstein's *Tractatus Logico-Philosophicus* for C.K. Ogden in 1922 from German into English, and one of his first scientific papers was a *Critical note of L. Wittgenstein's Tractatus Logico-Philosophicus* (1923).

John Maynard Keynes' influence on Ramsey took him into two subjects: probability and economics. In economics Ramsey wrote two papers only, viz. *A contribution to the theory of taxation* (1927) and *A mathematical theory of saving* (1928). But in his obituary of Ramsey Keynes called the latter "one of the most remarkable contributions to mathematical economics ever made" (see Braithwaite 1931).

Besides his profound work in mathematical logic and economy, Ramsey devoted a substantial part of his scientific life to philosophy. To cite Braithwaite (1931), a friend of Ramsey and latter professor of philosophy in Cambridge, once more: "though mathematical teaching was Ramsey's profession, philosophy was his

H.J. Prömel, *Ramsey Theory for Discrete Structures*,
DOI 10.1007/978-3-319-01315-2_1,
© Springer International Publishing Switzerland 2013

vocation". It is far beyond the scope of this attempt to survey Ramsey's work. The interested reader should consult, e.g., Ramsey (1978), for Ramsey's major works and for enlightening introductory comments, as well as Mellor (1983) for an excellent survey on Ramsey and his work.

Frank Plumpton Ramsey, Fellow of King's College, Cambridge, and University Lecturer in Mathematics died in 1930 at the age of 26.

1.2   Ramsey's Theorem

In 1928 Ramsey wrote a paper *On a problem in formal logic*, published in 1930 in the Proceedings of the London Mathematical Society for which he became eponymous for a part of discrete mathematics nowadays known as Ramsey theory.

The object was to give a decision procedure for propositional logic. The need for such procedures, we would say algorithms in the present day terminology, arose with the crisis of the foundations of mathematics around 1900. It was more or less the theory of sets and the arithmetization of analysis which led to this crisis. In response, the programs of Russell and Whitehead, of Hilbert and of Brouwer called for a new foundation trying to overwhelm the dubious principles of mathematics of that time.

It is a kind of irony that a purely mathematical result from Ramsey's paper proved to be of so much greater consequence than the metamathematical investigations for which they were made as tools. Even more, as it was discovered later, the full strength of Ramsey's theorem was not necessary to find a decision procedure for statements in the special class of first order logic investigated by Ramsey.

A few years afterwards two new proofs of Ramsey's theorem were obtained. By Skolem (1933) also applying this result to the decision problem of first order logic and by Erdős and Szekeres (1935) rediscovering Ramsey's theorem working on a problem in geometry (cf. Szekeres 1973).

We start discussing Ramsey's theorem by considering the countable infinite case. "The theorem which we actually require concerns finite classes only, but we shall begin with a similar theorem about infinite classes which is easier to prove and gives a simple example of the method of argument" (quoted from Ramsey 1930).

Theorem 1.1 (Ramsey).   *Let k and r be positive integers. Then for every r-coloring $\Delta : [\omega]^k \to r$ of the k-subsets of ω there exists an infinite subset $F \in [\omega]^\omega$ of ω such that all k-subsets of F have got the same color, i.e., $\Delta\!\restriction\![F]^k$ is a constant coloring.*

Proof. We prove Ramsey's theorem by induction on k, where the case $k = 1$ reduces to the pigeonhole principle. Assume the theorem is true for some $k \geq 1$ and let $\Delta : [\omega]^{1+k} \to r$ be a coloring.

Now assume that for some $j < \omega$ we have got already j elements $x_0 < \ldots < x_{j-1}$ and an infinite set $G \in [\omega]^\omega$ with $x_{j-1} < y = \min G$, such that $\Delta\!\restriction\!\{x_i\} \times [G]^k$

is constant for every $i < j$. We fix y. This induces a coloring $\Delta' : [G\backslash\{y\}]^k \to r$ of the k-subsets of $G\backslash\{y\}$ by $\Delta'(H) = \Delta(\{y\} \cup H)$. By the inductive assumption there is an $G' \in [G\backslash\{y\}]^\omega$ so that $\Delta'\!\restriction\![G']^k$ is constant. Hence, choosing $x_j = y$, we have that $\Delta\!\restriction\!\{x_i\} \times [G']^k$ is constant for every $i \leq j$. Continuing in this way we obtain an infinite set $X = \{x_i \mid i < \omega\}$ such that $\Delta(H) = \Delta(H')$, whenever $H, H' \in [X]^{1+k}$ satisfy $\min H = \min H'$. Applying the pigeonhole principle there exists $F \in [X]^\omega$ such that $\Delta\!\restriction\![F]^{1+k}$ is a constant coloring, completing the proof of Ramsey's theorem. □

To facilitate applications of Ramsey's theorem we introduce the arrow notation. For cardinals κ, λ, μ and ν the symbol $\kappa \to (\lambda)^\nu_\mu$ denotes the following partition property: for every partition of $[\kappa]^\nu$ into μ classes there exists a set of size λ whose ν-subsets are completely contained in one class. Using this notation Theorem 1.1 can be expressed by saying

$$\omega \to (\omega)^k_r, \quad \text{for all positive integers } k \text{ and } r.$$

This arrow notation was first used, in rudimentary form, in Erdős and Rado (1953). Applying König's lemma, a finite version of Theorem 1.1 can easily be obtained:

Theorem 1.2 (Ramsey). *Let k, m and r be positive integers. Then there exists a least positive integer $n = RAM(k, m, r)$ such that for every r-coloring $\Delta : [n]^k \to r$ of the k-subsets of $n = \{0, \ldots, n-1\}$ there exists an m-subset $M \in [n]^m$ of n such that $\Delta\!\restriction\![M]^k$ is a constant coloring.*

Proof. Assume that for every n there exists a coloring $\Delta : [n]^k \to r$ so that for every $M \in [n]^m$ we have that $\Delta\!\restriction\![M]^k$ is not constant. Such a Δ is called a bad coloring.

Obviously, the restriction $\Delta\!\restriction\![n-1]^k$ of any bad coloring $\Delta : [n]^k \to r$ is a bad coloring again. Hence, using the relation of being a restriction of, the bad colorings form a tree T, having the empty coloring as a root. T is locally finite and, by assumption, infinite. Therefore by König's lemma (König 1927), T contains an infinite path of bad colorings. This path defines a bad coloring of $[\omega]^k$ and, thus, contradicts Theorem 1.1. □

Ramsey, himself, did not use such a compactness argument to derive the finite version from the infinite one, but gave a quite elaborated explicit construction. Actually, the same idea as for the proof of Theorem 1.1 can be used to prove Theorem 1.2 directly, compare also Chap. 7.

1.3   Erdős-Szekeres' Theorem

One of the earliest and most popular applications of Ramsey's theorem is due to Erdős and Szekeres (1935). In fact, this application was a kind of first step in popularizing Ramsey's theorem also among non-logicians.

Theorem 1.3 (Erdős, Szekeres). *Let $m \geq 3$ be a positive integer. Then there exists a least positive integer $n = ES(m)$ such that any set of n points in the Euclidean plane, no three of which are collinear, contains m points which are the vertices of a convex m-gon.*

This result was conjectured (and proved in case $m = 4$) by E. Klein-Szekeres. The proof given here is from Johnson (1986).

Proof of Theorem 1.3. Choose n according to Ramsey's theorem such that $n \to (m)_2^3$ and let N be any set of n points in the plane, no three of which are collinear.

For $a, b, c \in N$ let $|abc|$ denote the number of points of N which lie in the interior of the triangle spanned by a, b and c. Now define $\Delta : [N]^3 \to 2$ by $\Delta(a, b, c) = 0$ if $|abc|$ is even and $\Delta(a, b, c) = 1$ otherwise. By choice of n there exists $M \subseteq [N]^m$ such that $\Delta\rceil[M]^3$ is constant. Then the points of M form a convex m-gon. Otherwise, there would be $a, b, c, d \in M$ so that d lies in the interior of the triangle abc. Since no three points of M are collinear we have

$$|abc| = |abd| + |acd| + |bcd| + 1,$$

contradicting that $\Delta\rceil[M]^3$ is constant.   □

In their 1935 paper Erdős and Szekeres give two proofs of this result. The second one, not using Ramsey's theorem, yields a smaller upper bound for $ES(m)$, viz. $ES(m) \leq \binom{2m-4}{m-2} + 1$. It can be shown that $2^{m-2} + 1 \leq ES(m)$ (Erdős and Szekeres 1961) and Erdős and Szekeres (1935) believe that $ES(m) = 2^{m-2} + 1$ is the correct value. Apart from some small numbers, this is still an open question.

1.4 Erdős-Rado's Canonization Theorem

The popularization of Ramsey's theorem is inherent with the names of Paul Erdős and Richard Rado. There are numerous of results, of each of them and of both which are basic in Ramsey theory. A good example is their joint paper *A combinatorial theorem* (Erdős and Rado 1950) which can be viewed as the first one in a part of Ramsey theory called canonizing Ramsey theory. This result is both: an application of Ramsey's theorem and a root for further development in Ramsey theory. The object of this theorem is to prove a generalization of Ramsey's theorem in which the number of colors need not to be finite.

Notation. Let $X \in [n]^k$, say $X = \{x_0, \ldots, x_{k-1}\}$ in ascending order, and let $J \subseteq k$. Then $X : J$ denotes the J-subset of X, i.e.,

$$X : J = \{x_j \mid j \in J\}.$$

Using this notation, the Erdős-Rado canonization theorem says:

Theorem 1.4 (Erdős-Rado canonization theorem). *Let k be a positive integer. Then for every coloring $\Delta : [\omega]^k \to \omega$ of the k-subsets of ω with arbitrary many colors there exists an infinite set $F \in [\omega]^\omega$ of ω and a (possibly empty) set $J \subseteq k$ such that*

$$\Delta(X) = \Delta(Y) \quad \text{if and only if} \quad X : J = Y : J,$$

holds for every $X, Y \in [F]^k$.

Following Erdős and Rado, we think it is worth while to state the case $k = 2$ explicitly:

Corollary 1.5. *For every coloring $\Delta : [\omega]^2 \to \omega$ of the pairs of ω with arbitrary many colors there exists an infinite set $F \subseteq \omega$ such that one of the following four conditions hold for all $X, Y \in [F]^2$, say $X = \{x_0, x_1\}$, $Y = \{y_0, y_1\}$ in ascending order:*

1. *$\Delta(X) = \Delta(Y)$,*
2. *$\Delta(X) = \Delta(Y)$ if and only if $x_0 = y_0$,*
3. *$\Delta(X) = \Delta(Y)$ if and only if $x_1 = y_1$,*
4. *$\Delta(X) = \Delta(Y)$ if and only if $x_0 = y_0$ and $x_1 = y_1$.*

Proof of Theorem 1.4. For a given coloring $\Delta : [\omega]^k \to [\omega]$, we choose $F \in [\omega]^\omega$ to be any set such that

(1) the patterns which Δ induces on $2k$-element subsets of F are all the same, i.e., $\Delta(X : I) = \Delta(X : J)$ if and only if $\Delta(Y : I) = \Delta(Y : J)$ for all $X, Y \in [F]^{2k}, I, J \in [2k]^k$.

To see that such F exists, set $r = \binom{2k}{k}$ and consider the coloring $\Delta' : [\omega]^{2k} \to r^{\binom{2k}{k}}$ defined as follows: for a set $X \in [\omega]^{2k}$, let $f : \{\Delta(X') \mid X' \subset X, |X'| = k\} \to r$ be any injective function and set $\Delta'(X) = \langle f(\Delta(X')) \mid X' \subset X, |X'| = k \rangle$. Ramsey's theorem guarantees an infinite monochromatic set F with respect to Δ', and it is easy to see that such set F satisfies property (1).

For the sake of brevity, we identify F in the following with ω. Further, for any set X and an integer s, we define

$$X_s = \{x \mid x \in X, \ x \le s\} \cup \{x + 1 \mid x \in X, \ x > s\}.$$

We will make use of (1) in two ways. Assume that $X = \{x_0, \ldots, x_{k-1}\}$, ordered ascendingly. Then for every $i < k$ we have (1a) that $\Delta(X_{x_i}) = \Delta(X_{x_i-1})$ if and only if $\Delta(\{0, \ldots, k\} \setminus \{i\}) = \Delta(\{0, \ldots, k\} \setminus \{i + 1\})$. Secondly, let $X = \{x_0, \ldots, x_{k-1}\}$ and $Y = \{y_0, \ldots, y_{k-1}\}$, again ordered ascendingly, and consider $Z = X \cup Y$. Assume I_x and I_y are such that $Z : I_x = X$ and $Z : I_y = Y$. Then, for any integer s, we have (1b) that $\Delta(X) = \Delta(Y)$ if and only if $\Delta(Z_s : I_x) = \Delta(Z_s : I_y)$.

We set $J = \{i < k \mid \Delta(\{0, \ldots, k\} \setminus \{i\}) \ne \Delta(\{0, \ldots, k\} \setminus \{i + 1\})\}$ and prove that this set J satisfies the property claimed in the theorem. To see this consider arbitrary sets $X = \{x_0, \ldots, x_{k-1}\}$ and $Y = \{y_0, \ldots, y_{k-1}\}$, ordered

ascendingly. First assume that $x_j = y_j$ for every $j \in J$. We show by induction
on $|\{i \mid x_i \neq y_i\}|$ that $\Delta(X) = \Delta(Y)$. Let $i = \max\{i \mid x_i \neq y_i\}$
and, say, $x_i < y_i$ ($< y_{i+1} = x_{i+1}$). Then $i \notin J$ and it thus follows
from (1) and the definition of J that $\Delta(\{x_0, \ldots, x_i, y_i, x_{i+1}, \ldots, x_{k-1}\} \backslash \{x_i\}) =$
$\Delta(\{x_0, \ldots, x_i, y_i, x_{i+1}, \ldots, x_{k-1}\} \backslash \{y_i\}) = \Delta(X)$. Thus, by induction, $\Delta(X) =$
$\Delta(Y)$.

Next assume that $\Delta(X) = \Delta(Y)$. We show that $x_j = y_j$ for every $j \in J$.
Suppose not, and let $j \in J$ be minimal with $x_j \neq y_j$, say, $x_j < y_j$. Let ℓ be
minimal such that $x_j \leq y_\ell$, and set $Z = X \cup Y$. Either $x_j < y_\ell$. Then by (1b) we
have $\Delta(Z_{x_j} : I_x) = \Delta(Z_{x_j} : I_y)$ and $\Delta(Z_{x_j-1} : I_x) = \Delta(Z_{x_j-1} : I_y)$. However,
by the choice of ℓ we have $Z_{x_j} : I_y = Z_{x_j-1} : I_y = \{y_0, \ldots, y_\ell+1, \ldots, y_{k-1}+1\}$.
Thus $\Delta(\{x_0, \ldots, x_j, x_{j+1}+1, \ldots, x_{k-1}+1\}) = \Delta(\{x_0, \ldots, x_j+1, \ldots, x_{k-1}+1\})$,
which is a contradiction to $j \in J$.

Or $x_j = y_\ell$. Then the minimality of j implies that $\ell \notin J$. Hence, using (1a) we
have $\Delta(Z_{y_\ell} : I_y) = \Delta(Z_{y_\ell-1} : I_y)$. Further, from (1b) we have $\Delta(Z_{x_j} : I_x) =$
$\Delta(Z_{x_j} : I_y) = \Delta(Z_{y_\ell} : I_y)$, as $y_\ell = x_j$, and also $\Delta(Z_{x_j-1} : I_x) = \Delta(Z_{y_\ell-1} :$
$I_y) = \Delta(Z_{y_\ell} : I_y)$. However, this implies $\Delta(Z_{x_j} : I_x) = \Delta(\{x_0, \ldots, x_j, x_{j+1} +$
$1, \ldots, x_{k-1} + 1\}) = \Delta(\{x_0, \ldots, x_j + 1, \ldots, x_{k-1} + 1\}) = \Delta(Z_{x_j-1} : I_x)$, again
contradicting $j \in J$.   □

A proof similar to this one is given in Rado (1986). Using König's lemma, a
finite version of the Erdős-Rado canonization theorem can easily be deduced from
Theorem 1.4:

Corollary 1.6. *Let k and m be positive integers. Then there exists a least positive
integer $n = ER(k, m)$ such that for every coloring $\Delta : [n]^k \to \omega$ of the k-subsets of
n with arbitrary many colors there exists an m-subset $M \in [n]^m$ of n and a (possibly
empty) set $J \subseteq k$ such that*

$$\Delta(X) = \Delta(Y) \quad \text{if and only if} \quad X : J = Y : J$$

holds for every $X, Y \in [M]^k$.   □

In fact, the same argument as in the proof of Theorem 1.4 can be used to proof
Corollary 1.6, relying on Theorem 1.5. In the next chapter we will also obtain it a
consequence of the canonical Graham-Rothschild theorem.

Chapter 2
From Hilbert's Cube Lemma to Rado's Thesis

Quite a while before Ramsey proved his partition theorem for finite sets some results have been established which can be viewed as the earliest roots of Ramsey theory. The probably first one is due to David Hilbert (1892). In connection with investigations on the irreducibility of rational functions with integer coefficients he proved that for every coloring of some sufficiently large interval $[1, n]$ with r colors, there exist positive integers $a, a_0, \ldots, a_{m-1} \leq n$ such that the affine m-cube

$$\{a + \sum_{i < m} \epsilon_i a_i \mid \epsilon_i \in \{0, 1\} \text{ for every } i < m\}$$

is completely contained in one color class. Apparently neither Hilbert himself nor some other mathematician at that time examined the underlying combinatorial principles of this lemma.

Others happened to a lemma proved by Issai Schur some 25 years later. In reproving a theorem of Dickson on a modular version of Fermat's conjecture, Schur (1916) showed that for every r-coloring of some sufficiently large interval $[1, n]$ there exist positive integers $a_0, a_1 \leq n$ such that the projective 2-cube

$$\{\sum_{i < 2} \epsilon_i a_i \mid \epsilon_i \in \{0, 1\} \text{ for every } i < 2\} \setminus \{0\}$$

is completely contained in one color class.

A conjecture of Schur concerning the distribution of quadratic residues, respectively nonresidues modulo p led Schur to a question on arithmetic progressions, which became famous as Baudet's conjecture (cf. Brauer 1973). The problem was solved by Bartel Leendart van der Waerden (1927). The corresponding theorem, well known as van der Waerden's theorem on arithmetic progressions, soon attracted many mathematicians. For example, Khinchin (1952) writing an elementary book on number theoretic problems selected this result as one of his *Three Pearls in Number Theory*.

H.J. Prömel, *Ramsey Theory for Discrete Structures*,
DOI 10.1007/978-3-319-01315-2__2,
© Springer International Publishing Switzerland 2013

Brauer (1928) used van der Waerden's theorem on arithmetic progressions to resolve Schur's conjecture on quadratic residues. In fact, Schur himself suggested a strengthening of van der Waerden's theorem which is in a sense a common generalization of Schur's lemma on projective 2-cubes and van der Waerden's theorem and allows to derive a stronger form of Schur's conjecture (Brauer 1928).

A first culmination point of Ramsey theory was obtained with the work of a student of Schur: Richard Rado. In a series of beautiful papers (Rado 1933a,b, 1943) based on his doctoral dissertation he extended the results of Hilbert, Schur and van der Waerden in a remarkable way. He gave among other results a complete characterization of all systems of homogeneous linear equations $\mathcal{L} = \mathcal{L}(x_0, \ldots, x_{m-1})$ over $\mathbb{Z}$ having the property that for every coloring of $\mathbb{Z}^+$ with finitely many colors, $\mathcal{L}$ has a monochromatic solution. Observe that Schur's lemma essentially says that $x + y = z$ has this property.

One convention: To avoid trivial cases we dismiss throughout this section the number 0. We consider colorings of $[1, n] = \{1, \ldots, n\}$ rather than $n = \{0, \ldots, n - 1\}$, and of $\mathbb{N}$, the set of positive integers, instead of ω.

2.1 Hilbert's Cube Lemma

Let a, m and $a_0, \ldots, a_{m-1}$ be positive integers. Then the set

$$\{a + \sum_{i < m} \epsilon_i a_i \mid \epsilon_i \in \{0, 1\} \text{ for every } i < m\}$$

is the affine m-cube generated by $a, a_0, \ldots, a_{m-1}$. Hilbert (1892) proved the following result:

Theorem 2.1 (Hilbert's cube lemma). *Let m and r be positive integers. Then for every r-coloring $\Delta : \mathbb{N} \to r$ of the positive integers there exists an affine m-cube which is monochromatic.*

Hilbert's cube lemma is probably the earliest result which can be viewed as a partition theorem (besides the pigeonhole principle, of course). It was established some 35 years before Ramsey's theorem. Hilbert's proof is written in the style of the late nineteeth century: detailed discussions appealing to the readers mathematical intuition. But despite its unusualness for todays reader the proof is convincing by its clarity and worth reading. So we think it is worth while to include the original proof of Hilbert (though in German). Later in this chapter (Sect. 2.3) we obtain Hilbert's lemma also from van der Waerden's theorem on arithmetic progressions.

Unsere Entwickelungen beruhen auf folgendem Hülfsatze:
    Es sei eine unendliche Zahlenreihe $a_1, a_2, a_3, \ldots$ vorgelegt, in welcher allgemein a_s eine der a ganzen positiven Zahlen $1, 2, \ldots, a$ bedeutet; es sei überdies m irgend eine ganze positive Zahl. Dann lassen sich stets m ganze positive Zahlen $\mu^{(1)}, \mu^{(2)}, \ldots, \mu^{(m)}$ so bestimmen, dass die 2^m Elemente

$$a_\mu,$$

$$a_{\mu+\mu^{(1)}},$$

$$a_{\mu+\mu^{(2)}}, a_{\mu+\mu^{(1)}+\mu^{(2)}},$$

$$a_{\mu+\mu^{(3)}}, a_{\mu+\mu^{(1)}+\mu^{(3)}}, a_{\mu+\mu^{(2)}+\mu^{(3)}}, a_{\mu+\mu^{(1)}+\mu^{(2)}+\mu^{(3)}},$$

$$\cdots\cdots\cdots$$

$$a_{\mu+\mu^{(\cdot\cdot)}}, a_{\mu+\mu^{(1)}+\mu^{(m)}}, a_{\mu+\mu^{(2)}+\mu^{(\cdot\cdot)}}, \ldots, a_{\mu+\mu^{(1)}+\mu^{(2)}+\ldots+\mu^{(m)}},$$

für unendlich viele ganzzahlige Werthe μ sämtlich gleich der nämlichen Zahl $\mathcal{G}$ sind, wo $\mathcal{G}$ eine der Zahlen $1, 2, \ldots, a$ bedeutet. Dabei wird der Index $\mu + \mu^{(1)}$ des zweiten Elementes erhalten, indem man die Zahl $\mu^{(1)}$ zu dem Index μ des ersten Elementes addirt; die Indices des dritten und vierten Elementes entstehen aus den Indices des ersten und zweiten Elementes, indem man zu diesen die Zahl $\mu^{(2)}$ addirt; die Indices des fünften, sechsten, siebenten, achten Elementes entstehen aus den Indices der vier ersten Elemente, wenn man zu diesen die Zahl $\mu^{(3)}$ addirt, und schliesslich erhält man die Indices der 2^{m-1} letzten Elemente, indem man zu den schon bestimmten Indices der 2^{m-1} ersten Elemente die Zahl $\mu^{(m)}$ addiert.

Beim Beweise ist es nothwendig, einzelne Theile der vorgelegten Reihe für sich zu betrachten. Wenn insbesondere i auf einander folgende Elemente der Reihe herausgegriffen werden, etwa die Elemente $a_\mu, a_{\mu+1}, a_{\mu+2}, \ldots, a_{\mu+i-1}$, so nenne ich diese i Elemente ein Intervall der Reihe von der Länge i. Wir grenzen nun innerhalb der vorgelegten Reihe irgend ein Intervall von der Länge $a + 1$ ab. In diesem Intervalle tritt dann mindestens eine der Zahlen $1, 2, \ldots, a$ etwa die Zahl $\mathcal{G}$, zweimal auf, d.h. in dem Intervalle von der Länge $a + 1$ kommt jedenfalls eine der folgenden Gruppirungen vor:

$$\mathcal{G}_2^{(1)} = \mathcal{G}\mathcal{G},$$

$$\mathcal{G}_3^{(1)} = \mathcal{G} \cdot \mathcal{G},$$

$$\mathcal{G}_4^{(1)} = \mathcal{G} \cdot \cdot \mathcal{G},$$

$$\cdots\cdots\cdots\cdots\cdots$$

$$\mathcal{G}_{a+1}^{(1)} = \mathcal{G} \cdots\cdots\cdots \mathcal{G}.$$

Wie schon durch die Schreibweise kenntlich gemacht ist, bedeutet hierin allgemein $\mathcal{G}_s^{(1)}$ ein Intervall von der Länge s, dessen erstes und letztes Element einander gleich, nämlich gleich der Zahl $\mathcal{G}$ sind. Man sieht, dass die Anzahl aller möglichen von einander verschiedenen Gruppirungen $\mathcal{G}_s^{(1)}$ gleich a^2 und somit jedenfalls kleiner als die Zahl $(a + 1)^2$ ist. Wir grenzen jetzt innerhalb der vorgelegten Reihe hinter einander $(a + 1)^2$ Intervalle ab, deren jedes die Länge $a + 1$ besitzt, und betrachten dann das so entstehende Gesammtintervall von der Länge $(a + 1)^3$. In demselben tritt nothwendig mindestens eine der Gruppirungen $\mathcal{G}_s^{(1)}$, etwa die Gruppirung $\mathcal{G}_{\nu^{(1)}}^{(1)}$, zweimal auf, d.h. in dem Intervalle von der Länge $(a + 1)^3$ kommt jedenfalls eine der folgenden Gruppirungen vor:

$$\mathcal{G}_{2\nu^{(1)}}^{(2)} = \mathcal{G}_{\nu^{(1)}}^{(1)} \mathcal{G}_{\nu^{(1)}}^{(1)},$$

$$\mathcal{G}_{2\nu^{(1)}+1}^{(2)} = \mathcal{G}_{\nu^{(1)}}^{(1)} \cdot \mathcal{G}_{\nu^{(1)}}^{(1)},$$

$$\mathcal{G}_{2\nu^{(1)}+2}^{(2)} = \mathcal{G}_{\nu^{(1)}}^{(1)} \cdot \cdot \mathcal{G}_{\nu^{(1)}}^{(1)},$$

$$\cdots\cdots\cdots\cdots\cdots$$

$$\mathcal{G}_{(a+1)^3}^{(2)} = \mathcal{G}_{\nu^{(1)}}^{(1)} \cdots\cdots\cdots \mathcal{G}_{\nu^{(1)}}^{(1)}.$$

Hier bedeutet allgemein $\mathcal{G}_s^{(2)}$ ein Intervall von der Länge s, welches mit der Gruppirung $\mathcal{G}_{\nu^{(1)}}^{(1)}$ beginnt und mit der nämlichen Gruppirung schliesst. Die Anzahl aller von einander verschiedenen Gruppirungen $\mathcal{G}^{(2)}$ ist offenbar kleiner als das Product der Intervalllänge $(a+1)^s$ in die Anzahl aller möglichen Gruppirungen $\mathcal{G}^{(1)}$, und folglich ist jene Anzahl der Gruppirungen $\mathcal{G}_s^{(2)}$ jedenfalls kleiner als $(a+1)^5$. Wenn wir daher innerhalb der vorgelegten Reihe hinter einander $(a+1)^5$ Intervalle abgrenzen und zwar ein jedes von der Länge $(a+1)^3$, so tritt in dem so entstehenden Intervalle von der Gesammtlänge $(a+1)^8$ mindestens eine der Gruppirungen $\mathcal{G}_s^{(2)}$, etwa die Gruppirung $\mathcal{G}_{\nu^{(2)}}^{(2)}$, zweimal auf, d. h. in dem Intervalle von der Länge $(a+1)^8$ kommt jedenfalls eine der folgenden Gruppirungen vor:

$$\mathcal{G}_{2\nu^{(2)}}^{(3)} = \mathcal{G}_{\nu^{(2)}}^{(2)} \mathcal{G}_{\nu^{(2)}}^{(2)},$$

$$\mathcal{G}_{2\nu^{(2)}+1}^{(3)} = \mathcal{G}_{\nu^{(2)}}^{(2)} \cdot \mathcal{G}_{\nu^{(2)}}^{(2)},$$

$$\mathcal{G}_{2\nu^{(2)}+2}^{(3)} = \mathcal{G}_{\nu^{(2)}}^{(2)} \cdot\cdot \mathcal{G}_{\nu^{(2)}}^{(2)},$$

$$\dots\dots\dots\dots$$

$$\mathcal{G}_{(a+1)^8}^{(3)} = \mathcal{G}_{\nu^{(2)}}^{(2)} \dots\dots\dots \mathcal{G}_{\nu^{(2)}}^{(2)}.$$

Hier bedeutet allgemein $\mathcal{G}_s^{(3)}$ ein Intervall von der Länge s, welches mit der Gruppirung $\mathcal{G}_{\nu^{(2)}}^{(2)}$ beginnt und mit der nämlichen Gruppirung schliesst.

Nach m-maliger Anwendung des nämlichen Verfahrens gelangen wir zu Gruppirungen von der Gestalt:

$$\mathcal{G}^{(m)} = \mathcal{G}^{(m-1)} \dots\dots\dots \mathcal{G}^{(m-1)}$$

und erkennen, dass in jedem Intervall der Reihe von einer gewissen Länge ℓ nothwendig eine jener Gruppirungen $\mathcal{G}^{(m)}$ vorkommen muss. Dabei bedeutet ℓ eine bestimmte endliche und nur von a und m abhängige Zahl. Die Anzahl aller von einander verschiedenen Gruppirungen $\mathcal{G}^{(m)}$ ergiebt sich wiederum kleiner als eine gewisse endliche Zahl k welche leicht aus a und m berechnet werden kann. In der vorgelegten Reihe können wir nun hinter einander beliebig viele Intervalle von der Länge abgrenzen, und es folgt daher, dass es unter den Gruppirungen $\mathcal{G}^{(m)}$ nothwendig eine giebt, welche in der vorgelegten Reihe unendlich oft vorkommt. Diese Gruppirung sei die folgende

$$\mathcal{G}_{\nu^{(m)}}^{(m)} = \mathcal{G}_{\nu^{(m-1)}}^{(m-1)} \dots\dots\dots \mathcal{G}_{\nu^{(m-1)}}^{(m-1)},$$

wo $\mathcal{G}_{\nu^{(m)}}^{(m)}$ und $\mathcal{G}_{\nu^{(m-1)}}^{(m-1)}$, Intervalle von der Länge $\nu^{(m)}$ beziehungsweise von der Länge $\nu^{(m-1)}$ bedeuten.

Wir erkennen hieraus leicht die Richtigkeit des obigen Hülfsatzes. Es ist nämlich die Gruppirung $\mathcal{G}_{\nu^{(m)}}^{(m)}$ durch die folgenden Recursionsformeln bestimmt:

$$\mathcal{G}_{\nu^{(1)}}^{(1)} = \mathcal{G} \dots\dots\dots \mathcal{G},$$

$$\mathcal{G}_{\nu^{(2)}}^{(2)} = \mathcal{G}_{\nu^{(1)}}^{(1)} \dots\dots\dots \mathcal{G}_{\nu^{(1)}}^{(1)},$$

$$\mathcal{G}_{\nu^{(3)}}^{(3)} = \mathcal{G}_{\nu^{(2)}}^{(2)} \dots\dots\dots \mathcal{G}_{\nu^{(2)}}^{(2)},$$

$$\dots\dots\dots\dots$$

$$\mathcal{G}_{\nu^{(m)}}^{(m)} = \mathcal{G}_{\nu^{(m-1)}}^{(m-1)} \dots\dots\dots \mathcal{G}_{\nu^{(m-1)}}^{(m-1)}.$$

wo stets die unteren Indices die Anzahl der Elemente angeben, aus denen die betreffenden Intervalle bestehen. Ich setze

$$\mu^{(1)} = v^{(1)} - 1,$$

$$\mu^{(2)} = v^{(2)} - v^{(1)},$$

$$\mu^{(3)} = v^{(3)} - v^{(2)},$$

$$\dots\dots\dots\dots$$

$$\mu^{(m)} = v^{(m)} - v^{(m-1)},$$

und behaupte dann, dass die so entstehenden, ganzen positiven Zahlen $\mu^{(1)}, \mu^{(2)}, \dots, \mu^{(m)}$ von derjenigen Beschaffenheit sind, welche unser Hülfsatz verlangt. In der That: es ist eben bewiesen worden, dass in der vorgelegten Reihe $a_1, a_2, a_3, \dots$ die Gruppirung $\mathcal{G}_{v^{(m)}}^{(m)}$ unendlich oft vorkommt, d. h. es giebt unendlich viele ganzzahlige Werthe von μ, für welche

$$a_\mu a_{\mu+1} \dots a_{\mu+v^{(m)}-1} = \mathcal{G}_{v^{(m)}}^{(m)}$$

wird. Aus dem Aufbau der Gruppirung $\mathcal{G}_{v^{(m)}}^{(m)}$ folgt dann

$$a_\mu = \mathcal{G},$$

$$a_{\mu+\mu^{(1)}} = \mathcal{G},$$

$$a_{\mu+\mu^{(2)}} = a_{\mu+\mu^{(1)}+\mu^{(2)}} = \mathcal{G},$$

$$a_{\mu+\mu^{(3)}} = a_{\mu+\mu^{(1)}+\mu^{(3)}} = a_{\mu+\mu^{(2)}+\mu^{(3)}} = a_{\mu+\mu^{(1)}+\mu^{(2)}+\mu^{(3)}} = \mathcal{G},$$

$$\dots\dots\dots\dots$$

$$a_{\mu+\mu^{(m)}} = a_{\mu+\mu^{(1)}+\mu^{(m)}} = a_{\mu+\mu^{(2)}+\mu^{(m)}} = \dots = a_{\mu+\mu^{(1)}+\mu^{(2)}+\dots+\mu^{(m)}} = \mathcal{G},$$

und damit ist der Hülfsatz bewiesen.

2.2 Schur's Lemma

Theorem 2.2 (Schur's lemma). *Let r be a positive integer. Then there exists a least positive integer $n = S(r)$, such that for every coloring $\Delta : [1, n] \to r$ there exist positive integers $x, y \le n$ satisfying*

$$\Delta(x) = \Delta(y) = \Delta(x + y).$$

Moreover, $S(r) \le er!$, where e is the base of the natural logarithm.

Let m and $a_0, \dots, a_{m-1}$ be positive integers. Then the set

$$\left\{ \sum_{i<m} \epsilon_i a_i \mid \epsilon_i \in \{0, 1\} \text{ for every } i < m \right\} \setminus \{0\}$$

is the projective m-cube generated by $a_0, \ldots, a_{m-1}$. Using this terminology Schur's lemma can be rephrased by saying that for every coloring $\Delta : [1, \lfloor er! \rfloor] \to r$ there exists a projective 2-cube which is monochromatic. Hence the lemma is a projective analogue to Hilbert's (affine) cube lemma for $m = 2$.

Hilbert used his lemma as a tool to obtain certain results on the irreducibility of rational functions with integer coefficients. Schur established his lemma to give an easy proof and moreover to extend a number theoretic theorem of Dickson showing that for each r the congruence

$$x^r + y^r \equiv z^r \pmod{p}$$

has solutions for all sufficiently large primes p.

We give two proofs of Schur's lemma. The first one follows the lines of Schur's original proof yielding the bound $S(r) \leq er!$ The second one is an application of Ramsey's theorem.

First Proof of Schur's lemma. Let $n_0 \geq er!$ and let $\Delta : [1, n_0] \to r$ be an r-coloring of the first n_0 positive integers. Assume that there do not exist integers $x, y \leq n_0$ such that $\Delta(x) = \Delta(y) = \Delta(x + y)$.

Let $i_0 < r$ be the color which occurs most frequently under the n_0 elements and let $\Delta^{-1}(i_0) = \{x_0, \ldots, x_{n_1-1}\}$ be in ascending order. Observe that $n_0 \leq rn_1$.

Consider $N_0 = \{x_i - x_0 \mid 1 \leq i < n_1\}$. By assumption $N_0 \cap \Delta^{-1}(i_0) = \emptyset$. Let i_1 be the most frequent color under the elements of N_0 and let $N_0 \cap \Delta^{-1}(i_1) = \{y_0, \ldots, y_{n_2-1}\}$ be in ascending order. Observe that $n_1 - 1 \leq (r - 1)n_2$.

Consider $N_1 = \{y_i - y_0 \mid 1 \leq i < n_2\}$. By assumption $N_1 \cap \Delta^{-1}(i_0) = \emptyset$ and $N_1 \cap \Delta^{-1}(i_1) = \emptyset$. Let i_2 be the most frequent color under the elements of N_1 and let $N_1 \cap \Delta^{-1}(i_2) = \{z_0, \ldots, z_{n_3-1}\}$ be in ascending order. Observe that $n_2 - 1 \leq (r - 2)n_3$.

Continue this procedure until some n_j becomes 1. At latest $n_r = 1$, as otherwise N_r contains two elements whose difference cannot not be colored by any of the r colors.

Inserting the above inequalities into each other gives eventually

$$n_0 \leq r! \cdot \left(1 + \sum_{i=1}^{r-1} \frac{1}{i!}\right) < r! \cdot e,$$

a contradiction to the choice of n_0. Hence, there exist x, y such that $\Delta(x) = \Delta(y) = \Delta(x + y)$. $\square$

Second proof of Schur's lemma. Let n be according to the finite Ramsey theorem such that $n \to (3)_r^2$ and let an r-coloring $\Delta : [1, n] \to r$ be given. This induces an r-coloring $\Delta^* : [n]^2 \to r$ by $\Delta^*(a, b) = \Delta(b - a)$ for $a < b$. By choice of n there exist $0 \leq u < v < w < n$ so that

$$\Delta^*(u, v) = \Delta^*(v, w) = \Delta^*(u, w)$$

and, hence, $\Delta(v - u) = \Delta(w - v) = \Delta(w - u)$. Putting $x = v - u$ and $y = w - v$ proves Schur's lemma.   □

Irving (1973) has slightly improved Schur's upper bound on $S(r)$ from $\lfloor r! e \rfloor$ to $\lfloor r!(e - \frac{1}{24}) \rfloor$. A lower bound is given in Fredricksen (1975, 1979), viz. $S(r) \geq c(315)^{\frac{r}{5}}$ for an appropriate constant c, cf. also Sect. 7.5.

2.3   Van der Waerden's Theorem

Schur, working on the distribution of quadratic residues and nonresidues, conjectured that for every k and every sufficiently large prime p there exist k consecutive numbers which are quadratic residues as well as k consecutive numbers which are quadratic nonresidues modulo p. To attack this conjecture he tried first to prove that for every k there exists n so that for every 2-coloring of $1, \ldots, n$ one of the two color classes contains an arithmetic progression of length k. He failed and both questions remained open for several years (cf. Brauer 1973).

Van der Waerden learned about the conjecture on arithmetic progressions most probably from P.J.H. Baudet at that time a young Dutch student in Göttingen. So his answer to this conjecture (van der Waerden 1927) is entitled *Beweis einer Baudetschen Vermutung*.

Theorem 2.3 (van der Waerden). *Let k and r be positive integers. Then there exists a least positive integer $n = W(k, r)$ such that for every r-coloring Δ : $[1, n] \to r$ there exists an arithmetic progression*

$$\{a + id \mid i < k\} \subseteq [1, n]$$

of length k which is monochromatic.

Years later, van der Waerden (1954, 1971) gave a personal account on *How the proof of Baudet's conjecture was found* – by now a classical contribution to the psychology of invention in mathematics.

Proof of Theorem 2.3. We prove actually something stronger than van der Waerden's theorem, namely:

Let k, m and r be positive integers. Then there exists a least positive integer $n = S(k, m, r)$ such that for every coloring Δ : $[1, n] \to r$ there exist positive integers a and $d_0, \ldots, d_{m-1}$ so that $a + k \cdot \sum_{i < m} d_i \leq n$ and

$$\Delta(a + \sum_{i < m} g_i d_i) = \Delta(a + \sum_{i < m} h_i d_i)$$

whenever $g, h \in (k + 1)^m$, where $g = (g_0, \ldots, g_{m-1})$ and $h = (h_0, \ldots, h_{m-1})$, agree up to their last occurrence of k (in g or h). Note: this implies that any combination of $g, h \in k^m$ is allowed, as then neither of them contains any k.

A set $\{a + \sum_{i<m} g_i d_i \mid g \in k^m\}$ is called *m-fold arithmetic progression*. Observe that for $m = 1$ we get the standard arithmetic progression of length k, thus $W(k, r) \leq S(k, 1, r)$ and van der Waerden's theorem follows.

We show the following two inequalities hold for all k, m, r:

1. $S(k, m+1, r) \leq S(k, m, r) \cdot S(k, 1, r^{S(k,m,r)})$,
2. $S(k+1, 1, r) \leq S(k, r, r)$

Together with the trivial observation that $S(1, 1, r) = 2$ for every r these inequalities immediately yield the proof of the statement by induction on m and k.

Proof of (1): Let $M = S(k, m, r)$ and $N = S(k, 1, r^{S(k,m,r)})$ and consider $\Delta :$ $[1, M \cdot N] \to r$. This induces a coloring $\Delta_N : [1, N] \to r^M$ by

$$\Delta_N(x) = \langle \Delta((x-1)M + j) \mid 1 \leq j \leq M \rangle.$$

By choice of N there exist positive integers b and d so that $\{b + jd \mid j < k\} \subseteq [1, N]$ and $\Delta_N \restriction \{b + jd \mid j < k\}$ is a constant coloring. Observe that this means that for any $1 \leq j \leq M$ we have

$$\Delta((b{-}1)M + j) = \Delta((b{-}1{+}d)M + j) = \ldots = \Delta((b{-}1{+}(k{-}1)d)M + j). \quad (2.1)$$

Next consider $\Delta_M : [(b-1)M + 1, bM] \to r$ where $\Delta_M = \Delta \restriction [(b-1)M + 1, bM]$. By choice of M there exist positive integers $a, d_0, \ldots, d_{m-1}$ so that the m-fold arithmetic progression $\{a + \sum_{i<m} g_i d_i \mid g \in (k+1)^m\}$ is completely contained in $[(b-1)M + 1, bM]$ and

$$\Delta(a + \sum_{i<m} g_i d_i) = \Delta(a + \sum_{i<m} h_i d_i) \quad (2.2)$$

for all $g, h \in (k+1)^m$ which agree up to their last occurrence of k. Let $d_m := dM$. We claim that then

$$\Delta(a + \sum_{i \leq m} g_i d_i) = \Delta(a + \sum_{i \leq m} h_i d_i)$$

for all $g, h \in (k+1)^{m+1}$ which agree up to their last occurrence of k. Note that the proof of this claim implies that (1) holds. In order to see why the claim holds observe first that if $g_m = k$ or $h_m = k$ then $g = h$ and the claim holds trivially. So assume $g_m, h_m < k$. Then the choice of $d_m = dM$ and (2.1) implies that

$$\Delta(a + \sum_{i \leq m} g_i d_i) = \Delta(a + \sum_{i < m} g_i d_i)$$

and

$$\Delta(a + \sum_{i \leq m} h_i d_i) = \Delta(a + \sum_{i < m} h_i d_i).$$

The claim thus follows immediately from (2.1).

Proof of (2): Let $N = S(k, r, r)$ and consider $\Delta : [1, N] \to r$. Then there exist $a, d_0, \ldots, d_{r-1}$ such that

$$\Delta(a + \textstyle\sum_{i<r} g_i d_i) = \Delta(a + \textstyle\sum_{i<r} h_i d_i), \tag{2.3}$$

whenever $g, h \in (k + 1)^r$ agree up to their last occurrence of k. Consider

$$\begin{aligned}
g^0 &= (\ 0\ ,\ 0\ ,\ \ldots\ ,\ 0\ ) \\
g^1 &= (\ k\ ,\ 0\ ,\ \ldots\ ,\ 0\ ) \\
&\ \vdots \\
g^r &= (\ k\ ,\ k\ ,\ \ldots\ ,\ k\ ).
\end{aligned}$$

By the pigeonhole principle two of these words, say g^μ and g^ν where $\mu < \nu$, have the property that $\Delta(a + \sum_{i<r} g_i^\mu d_i) = \Delta(a + \sum_{i<r} g_i^\nu d_i)$. More precisely,

$$\Delta(a + k \textstyle\sum_{i<\mu} d_i) = \Delta(a + k \textstyle\sum_{i<\mu} d_i + k \textstyle\sum_{i=\mu}^{\nu-1} d_i).$$

On the other hand, from (2.3) we have

$$\Delta(a + k \textstyle\sum_{i<\mu} d_i) = \Delta(a + k \textstyle\sum_{i<\mu} d_i + j \textstyle\sum_{i=\mu}^{\nu-1} d_i),$$

for every $j < k$. Thus, setting $a' = a + k \sum_{i<\mu} d_i$ and $d = \sum_{i=\mu}^{\nu-1} d_i$, we have that

$$\Delta \!\upharpoonright\! \{a' + j \cdot d \mid j < k + 1\}$$

is a constant coloring. Thus (2) holds.   □

The proof given above follows Graham and Rothschild (1974). Like the original proof of van der Waerden's theorem this proof also uses substantially that the assertion is known to be true for $k - 1$ and all r in order to derive it for k and some fixed r, say $r = 2$. Combinatorial proofs where the color number is fixed throughout the whole proof were obtained by Deuber (1982) using ideas from the proof of Hales-Jewett's theorem (cf. Sect. 4.1) and by Taylor (1982) giving a combinatorial version of the (Furstenberg and Weiss 1978) topological proof of van der Waerden's theorem.

The above proof has one disadvantage: the fact that it uses some kind of double induction yields an upper bound even on $W(k) := W(k, 2)$ which is not primitive recursive. In contrary to this the best lower bound currently available is $W(k + 1) \geq k2^k$, for k prime, which is due to Berlekamp (1968). Determining the order of magnitude of $W(k)$ or even proving that $W(k)$ increases slower than the Ackermann function has long been a challenging open problem in Ramsey theory. This was finally solved by Shelah (1988) who proved that the van der Waerden numbers are primitive recursive. The currently best asymptotic upper bound is by Gowers (2001). Some known exact values of $W(k)$ are $W(2) = 3$, $W(3) = 9$, $W(4) = 35$ and $W(5) = 178$ (see Chvátal 1970; Stevens and Shantaram 1978).

Let $H(m,r)$ denote the least integer for which the assertion of Hilbert's cube lemma (Theorem 2.1) is valid, i.e., the smallest number such that for every r-coloring $\Delta : [1, H(m,r)] \to r$ there exists an affine m-cube which is monochromatic with respect to Δ. Obviously, $H(m,r) \le W(m,r)$, as can be seen as follows. Let $\{a + jd \mid j < m\}$ be a monochromatic arithmetic progression. Then $a, d, d, \ldots, d$ (m many d's) generate a monochromatic affine m-cube proving Hilbert's cube lemma.

But in fact, $H(m,r)$ is much smaller than the van der Waerden number given above. Brown et al. (1985) showed using a result on B_2-sets that $H(2,r)$ is only quadratic in r, more precisely,

$$H(2,r) = (1 + o(1))r^2.$$

Moreover, examining Hilbert's original proof they observed that in general $H(m,r) \le r^{c^m}$ for an appropriate constant c. In other words, even for arbitrary m, the function $H(m,r)$ is bounded by a polynomial in r.

2.4   Schur's Extension of Van der Waerden's Theorem

"A few days" after van der Waerden answered Schur's question on arithmetic progressions, Brauer (1928) was able to use van der Waerden's result to resolve Schur's conjecture on quadratic residues and nonresidues. But Brauer's paper contains also a strengthening of van der Waerden's theorem (and of Schur's lemma) which he attributes to Schur (cf. also Brauer 1973):

Theorem 2.4. *Let k and r be positive integers. Then there exists a least positive integer $n = SB(k,r)$ such that for every r-coloring $\Delta : [1,n] \to r$ there exists an arithmetic progression*

$$\{a + id \mid i < k\} \subseteq [1,n]$$

of length k which is monochromatic and its difference d is in the same color, i.e., $\Delta \rceil (\{a + id \mid i < k\} \cup \{d\})$ is a constant coloring.

Proof. We proceed by induction on the color number r, the case $r = 1$ being trivial for every k.

Assume that the existence of $SB(k, r - 1)$ has been established for some $r > 1$. Choose $n = W(k \cdot SB(k, r - 1) + 1, r)$ and let $\Delta : [1, n] \to r$ be an arbitrary r-coloring. By choice of n there exists an arithmetic progression

$$\{a + jd' \mid j \le k \cdot SB(k, r - 1)\}$$

which is monochromatic with respect to Δ, say in color $r - 1$.

Now either for some j, $0 < j \leq SB(k, r-1)$, we have $\Delta(jd') = r - 1$. In this case we are done with a and $d = jd'$. Or $\Delta \upharpoonright \{jd' \mid 0 < j \leq SB(k, r-1)\}$ is an $(r-1)$-coloring. In that case using the inductive hypothesis finishes the proof. □

We outline the proof of Schur's conjecture using this strengthening of van der Waerden's theorem.

Let p be a prime number and let n be prime to p. Recall that n is a quadratic residue modulo p if $x^2 \equiv n \pmod{p}$ for some positive integer x; otherwise n is a quadratic nonresidue modulo p. Thus the set of integers is divided into three classes, the class of quadratic residues, the class of quadratic nonresidues and the multiples of p. The Legendre symbol $\left(\frac{n}{p}\right)$ is used to indicate the quadratic character of a number. Its value is ± 1 according to whether n is (or is not) a quadratic residue modulo p. There exist $\frac{1}{2}(p-1)$ quadratic residues, respectively, $\frac{1}{2}(p-1)$ non-residues modulo p in $\mathbb{Z}_p$.

Theorem 2.5. *Let k be a positive integer. Then there exists a positive integer $n = n(k)$ such that for every prime number $p > n$ there exist k consecutive integers which are quadratic residues modulo p and there exist k consecutive integers which are quadratic nonresidues modulo p.*

Proof. First we show that for every sufficiently large prime there exist k consecutive integers which are quadratic residues modulo p.

Let $n = SB(k, 2)$ and $p > n$ be a prime number. Color $[1, \ p-1]$ according to being a quadratic residue modulo p. By choice of n there exists an arithmetic progression

$$\{a + jd \mid j < k\} \subseteq [1, \ p-1]$$

which is monochromatic and its difference d is in the same color, i.e., the Legendre symbol $\left(\frac{x}{p}\right)$ is constant on $\{a + jd \mid j < k\} \cup \{d\}$.

As the product of two quadratic residues as well as the product of two quadratic nonresidues are quadratic residues, whereas the product of a quadratic residue and a quadratic nonresidue is a nonresidue we deduce that

$$\{\frac{a + jd}{d} \mid j < k\}$$

(with division in the Gallois field $\mathbb{Z}_p$) is a sequence of k consecutive quadratic residues modulo p proving the first part of the theorem.

Now let $\ell = (k! - 1)(k - 1) + 1, n = SB(\ell, 2)$ and $p > n$ be a prime number. According to the first part of the proof there exists a sequence

$$\{b + j \mid j < \ell\} \subseteq [1, p-1]$$

of ℓ consecutive quadratic residues modulo p. Let d be the smallest nonresidue modulo p. If $d < k!$ then

$$\{\frac{b + jd}{d} \mid j < k\}$$

(with division again in the field $\mathbb{Z}_p$) is a sequence of k consecutive nonresidues. So we can assume that $d \geq k!$. But then $d = k!m + c$ where $c < k!$. Therefore $c - d \equiv 0 \pmod{j}$ and hence $c - d + jd \equiv 0 \pmod{j}$ for every $0 < j < k$. But this gives that

$$\frac{c - d}{j} + d < d$$

and so by assumption $\frac{c-d}{j} + d$ is a quadratic residue. Since $j < k \leq d$, also $(\frac{c-d}{j} + d)j$ is a quadratic residue. Therefore

$$\{(c - d) + jd \mid j < k\}$$

is a progression of quadratic residues but its difference d is a nonresidue. Dividing by d yields the desired result.   □

2.5 Rado's Thesis

2.5.1 Partition Regular Systems of Homogenous Linear Equations

Let $Ax = 0$ be a system of homogenous linear equations in n variables with integer coefficients. Then $Ax = 0$ is *partition regular* if for every coloring of the positive integers with finitely many colors there exists a monochromatic solution, in other words, there exist positive integers $x_0, \ldots, x_{n-1}$ (not necessarily distinct) so that $A(x_0, \ldots, x_{n-1})^T = 0$ and $x_0, \ldots, x_{n-1}$ are all in the same color.

Schur's lemma asserts that

$$x_0 + x_1 - x_2 = 0$$

is partition regular, Schur's extension of van der Waerden's theorem that for every k the system

$$x_1 = x_0 + d$$
$$x_2 = x_1 + d$$

$$\vdots$$

$$x_k = x_{k-1} + d$$

is partition regular and Hilbert's cube lemma implies that

$$a + \sum_{i \in I} x_i = x_I, \quad I \subseteq n, \ I \neq \emptyset$$

is a partition regular system of equations.

Observe that using a compactness argument as in the proof of the finite Ramsey theorem (Theorem 1.2) it follows immediately that if $Ax = 0$ is partition regular then for each positive integer r there exists already a positive integer $N = N(A, r)$ such that for every r-coloring of $[1, N]$ there exists a monochromatic solution of $Ax = 0$ in $[1, N]$.

The notion of partition regularity is defined only for positive integers and all examples considered so far deal only with colorings of positive integers. One might think that additional linear systems of equations turn out to be partition regular if we consider r-colorings of nonzero rationals. The following lemma shows that this is not the case.

Lemma 2.6. *Let A be a matrix with integer coefficients. Then the following properties are equivalent:*

(1) *$Ax = 0$ is partition regular,*
(2) *For every coloring of the non-zero integers with finitely many colors there exists a monochromatic solution of $Ax = 0$,*
(3) *For every coloring of the non-zero rationals with finitely many colors there exists a monochromatic solution of $Ax = 0$.*

Proof. Since $\mathbb{N} \subseteq \mathbb{Z}\backslash\{0\} \subseteq \mathbb{Q}\backslash\{0\}$, we have trivially the implications from (1) to (2) and from (2) to (3).

Assume (3) and let r be a positive integer. By a compactness argument (König's lemma) there exists a finite set $S \subseteq \mathbb{Q}\backslash\{0\}$ such that for every r-coloring of S there exists a monochromatic solution of $Ax = 0$ in S. Multiply S with an appropriate integer c such that $\{cs \mid s \in S\} \subseteq \mathbb{Z}\backslash\{0\}$. Then for every r-coloring of $\{cs \mid s \in S\}$ there exists a monochromatic solution of $Ax = 0$, showing (2).

Now assume (2) and let $\Delta : \mathbb{N} \to r$ be a coloring. Define $\Delta' : \mathbb{Z}\backslash\{0\} \to 2r$ by

$$\Delta'(z) = \begin{cases} \Delta(z) & \text{if } z > 0 \\ \Delta(-z) + r & \text{if } z < 0. \end{cases}$$

Then by homogeneity the required result follows. □

Observe that by homogeneity we could also replace A by a matrix with rational coefficients.

Based on his thesis written under the supervision of Schur, Rado provided a complete characterization of all systems of homogenous linear equations which are partition regular. The crucial notion in this characterization is the column property of a matrix.

Definition 2.7. Let A be a matrix with integer coefficients, say $A = (a^0, \ldots, a^{n-1})$ where the a^i are the columns of A. Then A has the *column property* if there exists a partition of n, say $n = I_0 \cup \ldots \cup I_\ell$ for some $\ell < n$, such that

1. $\displaystyle\sum_{i \in I_0} a^i = 0$, i.e., the columns in I_0 add up to $\mathbf{0}$, and

2. for every $j < \ell$ there exist rational numbers ξ_{ij} such that

$$\sum_{i \in I_{j+1}} a^i = \sum_{i \in \bigcup_{v \le j} I_v} \xi_{ij} a^i,$$

i.e., the sum of the columns in I_{j+1} is a rational linear combination of the columns in the previous classes $I_0 \cup \ldots \cup I_j$.

We now consider some examples.

(1) The matrix

$$(1, 1, -1)$$

which describes Schur's equation $x_0 + x_1 = x_2$ obviously has the column property.

(2) The matrix of the system of equations

$$x_{i+1} = x_i + d, \quad i < k$$

has the column property choosing I_0 and I_1 as depicted below:

$$\begin{pmatrix} 1 & 1 & -1 & 0 & \cdots & 0 \\ 1 & 0 & 1 & -1 & \cdots & 0 \\ \vdots & \vdots & & & \ddots & \\ 1 & 0 & \cdots & 0 & 1 & -1 \end{pmatrix}$$
$$\underbrace{}_{I_1} \underbrace{}_{I_0}$$

(3) As a third example we consider the matrix corresponding to the system

$$\sum_{i \in I} x_i = x_I, \quad I \subseteq n, I \ne \emptyset,$$

which is a projective version of the system of equations we get from Hilbert's cube lemma. For $n = 2$ we obtain the matrix of Example (1) where the trivial equations are omitted. For $n = 3 = \{0, 1, 2\}$ the corresponding matrix is given below

$$\begin{pmatrix} 1 & 1 & 1 & -1 & 0 & \cdots & 0 \\ 1 & 1 & 0 & 0 & -1 & \cdots & 0 \\ 1 & 0 & 1 & & & & \\ 1 & 0 & 0 & & & & \\ 0 & 1 & 1 & & & & \\ 0 & 1 & 0 & & & & \\ 0 & 0 & 1 & 0 & \cdots & \cdots & -1 \end{pmatrix}$$

It can easily be seen that such matrices have the column property: assume that the matrix is arranged so that the rows are ordered lexicographically from the top to the bottom with respect to $1 > 0$ and the ith row contains exactly one -1 positioned in the $(n + i)$th column (compare the picture above). Then

$$I_0 = \{a^0\} \cup \{a^n, \ldots, a^{n+2^{n-1}-1}\}, \text{ and}$$

$$I_j = \{a^j\} \cup \{a^{n+\sum_{i=1}^{j} 2^{n-i}}, \ldots, a^{n+\sum_{i=1}^{j+1} 2^{n-i}-1}\} \quad \text{for } 1 \le j < n,$$

gives the desired column partition.

Theorem 2.8 (Rado). *Let A be a matrix with integer coefficients. Then the homogeneous system $Ax = 0$ of linear equations is partition regular if and only if the matrix A has the column property.*

We postpone the proof of Rado's theorem until we have introduced the so-called (m, p, c)-sets which can be viewed as generalizations of arithmetic progressions. The notion of (m, p, c)-sets was invented by W. Deuber in his doctoral dissertation where he proved a partition theorem for these sets as a tool to answer a long standing conjecture of Rado in the affirmative (Deuber 1973).

We will use this partition theorem for (m, p, c)-sets to prove Rado's theorem.

2.5.2 (m, p, c)-Sets

Definition 2.9. Let m, p, c be positive integers. A set $M \subseteq \mathbb{N}$ is an (m, p, c)-set if there exist positive integers $x_0, \ldots, x_m$ such that

$$M = M_{p,c}(x_0, \ldots, x_m)$$

$$= \{cx_i + \sum_{j=i+1}^{m} \xi_j x_j \mid \xi_j \in [-p, p] \cap \mathbb{Z} \text{ for every } j \in [i + 1, m] \text{ and } i \le m\}.$$

Observe that a $(1, k, 1)$-set is an arithmetic progression together with its difference and an $(n, 1, 1)$-set contains solutions to the system of equations given in Example (3) in the last section. Intuitively speaking, (m, p, c)-sets are m-fold arithmetic progressions together with c-fold differences.

We show that every system $Ax = 0$ of homogeneous linear equations given by a matrix A having the column property admits to find positive integers m, p and c such that every (m, p, c)-set contains a solution of $Ax = 0$.

Together with a partition theorem for (m, p, c)-sets this will yield that $Ax = 0$ is partition regular.

Lemma 2.10. *Let A be a matrix with integer coefficients having the column property. Then there exist positive integers m, p and c such that every (m, p, c)-set contains a solution of $Ax = 0$.*

Proof. Let $A = (a^0, \ldots, a^{n-1})$. By definition there exists a partition $n = I_0 \cup \ldots \cup I_\ell$ such that $\sum_{i \in I_0} a^i = 0$ and for every $j < \ell$ there exist rationals ξ_{ij} so that

$$\sum_{i \in I_{j+1}} a^i = \sum_{i \in \bigcup_{v \leq j} I_v} \xi_{ij} a^i .$$

Put $m = \ell + 1$ and let c be the least common multiple of the denominators of the ξ_{ij}. Finally, define $\tilde{p}$ to be the maximum of the absolute values of the ξ_{ij} and put $p = \tilde{p} \cdot c$. We claim that m, p and c have the desired properties.

We now show by induction on k that every (k, p, c)-set contains a solution of the matrix A_k consisting of those columns of A belonging to the classes $\bigcup_{i \leq k} I_i$.

Clearly, this is true for the matrix $A_0 = (a^i \mid i \in I_0)$ since every singleton provides a solution of $A_0 x = 0$. Now consider the $(k + 1, p, c)$-set $M = M_{p,c}(x_0, \ldots, x_{k+1})$ for some $k \geq 0$. By induction hypothesis we know that the (k, p, c)-set $M_{p,c}(x_0, \ldots, x_k) \subseteq M$ contains a solution of $A_k = (a^i \mid i \in \bigcup_{j \leq k} I_j)$, say

$$\sum_{i \in \bigcup_{j \leq k} I_j} y_i a^i = 0, \quad \text{where } y_i \in M_{p,c}(x_0, \ldots, x_k) \text{ for every } i .$$

By the column property of A and by choice of p there exist integers ξ_{ik}^c with $|\xi_{ik}^c| \leq p$ so that

$$\sum_{i \in \bigcup_{j \leq k} I_j} \xi_{ik}^c a^i + c \sum_{i \in I_{k+1}} a^i = 0 .$$

Multiplying this equation with x_{k+1} and adding it to the first one yields

$$\sum_{i \in \bigcup_{j \leq k} I_j} (\xi_{ik}^c x_{k+1} + y_i) a^i + \sum_{i \in I_{k+1}} c x_{k+1} a^i = 0 .$$

Recall that $y_i \in M_{p,c}(x_0, \ldots, x_k)$ for every i and $|\xi_{ik}^c| \leq p$. Hence, $y_i + \xi_{ik}^c x_{k+1} \in M_{p,c}(x_0, \ldots, x_{k+1})$. Obviously, $c x_{k+1} \in M_{p,c}(x_0, \ldots, x_{k+1})$. Thus we have constructed a solution of $A_{k+1} x = 0$ which is contained in $M_{p,c}(x_0, \ldots, x_{k+1})$. $\qquad\square$

The following partition theorem for (m, p, c)-sets is from Deuber (1973):

Theorem 2.11 (Deuber). *Let m, p, c and r be positive integers. Then there exist positive integers n, q and d such that for every coloring $\Delta : \mathbb{N} \to r$ of the positive integers every (n, q, d)-set $N \subseteq \mathbb{N}$ contains a monochromatic (m, p, c)-set.*

Combining this result with Lemma 2.10 proves the partition theoretic part of Rado's theorem, viz. that A having the column property implies that $Ax = 0$ is partition regular. In fact Theorem 2.11 is stronger than needed for our purposes. Deuber used this full partition theorem for (m, p, c)-sets to answer a conjecture of Rado:

A subset $S \subseteq \mathbb{N}$ is called partition regular if every partition regular system of equations is solvable in S. Deuber showed that if S is partition regular and S is colored with finitely many colors then one of the color classes is again partition regular.

Originally, Theorem 2.11 was proved with the help of van der Waerden's theorem on arithmetic progressions. Later, Leeb (1975) observed that the use of Hales-Jewett's theorem provides a more elegant proof.

A proof of Deuber's theorem based on Hales-Jewett's theorem will be given in Sect. 4.2.

2.5.3 Proof of Rado's Theorem

Deuber's Theorem 2.11 together with Lemma 2.10 implies that the column property of A implies that $Ax = 0$ is partition regular. Hence, it remains to show that the partition regularity of $Ax = 0$ implies the column property of A.

Let $A = (a^0, \ldots, a^{n-1})$ be a $k \times n$-matrix such that $Ax = 0$ is partition regular.

Let $I \subseteq n$ and $a \neq 0$ be a vector in $\mathbb{Z}^k$ that is not a (rational) linear combination of the $a^i, i \in I$. Let $P(I, a)$ be the set of all primes p such that for some nonnegative integer m we have that $p^m \cdot a$ is a linear combination of the $a^i, i \in I$, modulo p^{m+1}. Then $P(I, a)$ is finite.

To see this let $b \in \mathbb{Q}^k$ be such that $b^T \cdot a^i = 0$ for every $i \in I$ but $b^T \cdot a \neq 0$. Without loss of generality we can assume that $b \in \mathbb{Z}^k$ and, hence, $b^T \cdot a \in \mathbb{Z}$.

Let m be some nonnegative integer. Then

$$p^m a = \sum_{i \in I} \xi_i a^i \quad (\text{mod } p^{m+1})$$

implies that

$$b^T p^m a = 0 \pmod{p^{m+1}}.$$

Hence, $p \mid b^T \cdot a$ which is only true for finitely many primes.

Now choose a prime p which is not in $P(I, a)$ for every $a = \sum_{j \in J} a^j$ where $J \subseteq n$ and a is not a linear combination of $a^i, i \in I$. Moreover, let p be not one of the finitely many primes which have the property that $\sum_{i \in I} a^i \equiv 0 \pmod{p}$ for some $I \subseteq n$ with $\sum_{i \in I} a^i \neq 0$.

Every positive integer x admits a unique representation as $x = y(x) p^{z(x)}$ where $y(x) \not\equiv 0 \pmod{p}$. Let $\Delta_p : \mathbb{N} \to [1, p-1]$ be the coloring given by $\Delta_p(x) = y(x) \pmod{p}$. Since $Ax = 0$ is partition regular there exists a solution which is monochromatic with respect to Δ_p. This solution has the form

$$x_i = p^{z(x_i)}(p\alpha(x_i) + r) \quad \text{for every } i < n,$$

where $r \in [1, p-1]$ is the same for every i. Without loss of generality we can assume that

$$z(x_0) \leq \ldots \leq z(x_{n-1}).$$

We will partition n according to the $z(x_i)$-values and show that this partition proves that A has the column property. For this purpose let

$$m_0 = z(x_0) = \ldots = z(x_{i_1})$$
$$m_1 = z(x_{i_1+1}) = \ldots = z(x_{i_2})$$

$$\vdots$$

$$m_\ell = z(x_\ell + 1) = \ldots = z(x_{n-1}) \text{ and}$$
$$m_0 < m_1 < \ldots < m_\ell$$

Now put

$$I_0 = \{0, \ldots, i_1\}$$
$$I_1 = \{i_1 + 1, \ldots, i_2\}$$

$$\vdots$$

$$I_\ell = \{i_t + 1, \ldots, n-1\}.$$

First we verify that $\sum_{i \in I_0} a^i = 0$. Since $x_0, \ldots, x_{n-1}$ is a solution we have that

$$\sum_{i < n} x_i a^i = 0.$$

Thus in particular

$$\sum_{i \in I_0} x_i a^i + \sum_{i \in n \setminus I_0} x_i a^i \equiv 0 \pmod{p^{m_0 + 1}}.$$

For $i \in n \setminus I_0$ we have that $x_i \equiv 0 \pmod{p^{m_0 + 1}}$ and for every $i \in I_0$ that $x_i = p^{m_0}(p\alpha(x_i) + r)$. Hence

$$r \cdot \sum_{i \in I_0} a^i \equiv 0 \pmod{p}.$$

Since $r \in [1, p-1]$ and by choice of p it follows that $\sum_{i \in I_0} a^i = 0$.

Now we verify along the same lines that for $k > 0$ the sum of the columns in class I_k is a linear combination of the columns in the previous classes. As before, we have that

$$\sum_{i \in \bigcup_{j < k} I_j} x_i a^i + \sum_{i \in I_k} x_i a^i + \sum_{\in \bigcup_{k < j} I_j} x_i a^i \equiv 0 \pmod{p^{m_k + 1}}.$$

Hence, reducing modulo p gives

$$\sum_{i \in \bigcup_{j < k} I_j} x_i a^i + r p^{m_k} \sum_{i \in I_k} a^i \equiv 0 \pmod{p^{m_k + 1}}.$$

Thus by choice of p we obtain the desired result, completing the proof of Rado's theorem. □

It should be mentioned that Furstenberg (1981) obtained a proof of Rado's theorem of completely different nature using methods from topological dynamics.

2.5.4 Finite and Infinite Sums

Of course, Rado's theorem covers Hilbert's cube lemma, Schur's lemma and van der Waerden's theorem as well as Schur's extension of it. Because of its particular interest we will briefly discuss one other special case of Rado's theorem.

Recalling Example (3) we get as an immediate consequence of Rado's theorem the following finite sum theorem:

Theorem 2.12 (Rado, Folkman, Sanders). *Let m and r be positive integers. Then there exists a least positive integer $n = FS(m, r)$ such that for every coloring $\Delta : [1, n] \to r$ there are m positive integers $a_0, \ldots, a_{m-1}$ such that for all nonempty sets $I, J \subseteq m$ it follows that*

$$\Delta(\sum_{i \in I} a_i) = \Delta(\sum_{j \in J} a_j).$$

Theorem 2.12 was rediscovered several times, among others by Folkman (see Graham 1981 or Graham et al. 1980) and Sanders (1968) leading to the present name of this theorem.

Folkman's proof uses van der Waerden's theorem. The idea of the second proof of Schur's lemma (cf. Sect. 2.2) has been extended by Nešetřil and Rödl (1983a) to obtain a proof of the Rado-Folkman-Sanders theorem from Ramsey's theorem. In Sect. 5.2.4 we will get the finite sum theorem as an immediate application of Hales-Jewett's theorem.

Another combinatorial proof of the finite sum theorem was given by Taylor (1981). His proof is remarkable because it provides the least known upper bound on $FS(m, r)$, viz.

$$\text{FS} \leq 2^{r^{3^{r^{3^{\cdot^{\cdot^{\cdot^{r^3}}}}}}}} \Big\} 2r(m-1) \quad , m, r \geq 2.$$

Having the finite sum theorem in hands it is natural to ask whether or not an infinite version of it is valid. Graham and Rothschild (1971) conjectured an infinite generalization of the Rado-Folkman-Sanders theorem which was proved by Hindman (1974):

Theorem 2.13 (Hindman). *Let r be a positive integer. Then for every coloring $\Delta : \mathbb{N} \to r$ there exist infinitely many integers $a_0, a_1, a_2, \ldots$ such that for all nonempty finite sets $I, J \subseteq \omega$ it follows that*

$$\Delta(\sum_{i \in I} a_i) = \Delta(\sum_{j \in J} a_j),$$

i.e., all finite sums of the a_i get the same color.

Several proofs have been given for this theorem, e.g., by Baumgartner (1974) using some kind of combinatorial forcing, by Glazer (see, Hindman 1979) using idempotent ultrafilters in $\beta\mathbb{N}$ and by Furstenberg and Weiss (1978) using topological dynamics. The reader may consult one of these references for a proof of Hindman's theorem.

Assuming the axiom of choice it is easy to see that (coloring the reals) one cannot expect to get also infinite sums in the same color. Of course, taking infinite sums necessarily requires convergence. But restricting to in a sense constructive colorings, viz. colorings having the property that each color class has the property of Baire, it can be proved that there always exists an infinite sequence of reals (whose sum converges) such that all their sum (finite or infinite, but without repetition) get the same color (Prömel and Voigt 1990).

Part II
A Starting Point of Ramsey Theory: Parameter Sets

Chapter 3
Definitions and Basic Examples

In their by now classical paper *Ramsey's theorem for n-parameter sets* Graham and Rothschild (1971) introduced the concept of parameter sets. The idea was to find a combinatorial abstraction of linear and affine vector spaces over finite fields. This was motivated by a conjecture of Rota, proposing a geometric analogue to Ramsey's theorem. In fact, the Ramsey theorem for n-parameter sets implies Rota's conjecture directly for lower dimensional cases and, as it has turned out, the method used in the proof of this theorem contains also the seeds of the ideas to prove Rota's conjecture in its full strength. This was done in Graham, Leeb and Rothschild (1972).

But the impact of parameter sets goes far beyond the proof of Rota's conjecture. For example, Ramsey's theorem itself is an immediate consequence of the Graham-Rothschild theorem.

In a more rudimentary form parameter sets occur already in the paper *Regularity and positional games* by Hales and Jewett, published in 1963, who proved in a sense a pigeon hole principle for parameter sets.

The theorem of Hales and Jewett revealed the combinatorial core of van der Waerden's theorem on arithmetic progressions. But the concept of parameter words does not only glue together arithmetic progressions and finite sets. It allows a unifying approach to several seemingly different structures like Boolean lattices, Partition lattices, hypergraphs, and Deuber's (m, p, c)-sets, just to mention a few.

To a certain extend the Graham-Rothschild theorem can be viewed as a starting point of Ramsey theory.

Besides the various applications, several ramifications and generalizations of the original Graham-Rothschild theorem have been discovered. In this chapter we discuss the origins and some developments based on and related to the structure of Graham-Rothschild parameter sets.

H.J. Prömel, *Ramsey Theory for Discrete Structures*,
DOI 10.1007/978-3-319-01315-2_3,
© Springer International Publishing Switzerland 2013

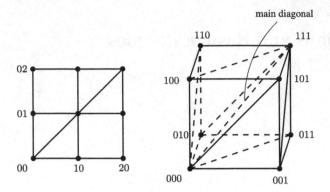

Fig. 3.1 The lines of A^2 when $A = 3$ (*left*) and the lines of A^3 when $A = 2$ (*right*)

3.1   Parameter Words

Unless stated otherwise, A is a finite set (alphabet). We are concerned with A^n, the set of n-tuples over A, certain subsets of this set, parameter sets, and their representations, parameter word.

Zero-parameter sets are simply singleton elements of A^n. A one-parameter set (or *combinatorial line*) $L \subseteq A^n$ is a set of size $|A|$ such that there exists a nonempty set $I \subseteq n$ of coordinates and for every $i \in n \backslash I$ there exists an element $a_i \in A$ such that

$$L = \{(x_0, \ldots, x_{n-1}) \mid x_i = x_j \text{ for all } i, j \in I \text{ and } x_i = a_i \in A \text{ for } i \notin I\}.$$

Intuitively speaking the set I consists of the moving coordinates and the coordinates $i \in n \backslash I$ that are constant.

Obviously, putting $t = |A|$ there are $(t + 1)^n - t^n$ lines in A^n. As examples, in Fig. 3.1 we indicate the lines of A^2 when $A = 3 = \{0, 1, 2\}$ and the lines of A^3 when $A = 2 = \{0, 1\}$.

Every one-parameter set can be represented by a one-parameter word $f \in (A \cup \{\lambda_0\})^n$, containing the parameter λ_0 at least once, such that L results from f by replacing λ_0 by elements of A. Thus the parameter λ_0 indicates the moving coordinates. For example, $L = \{(0, 1, 0), (1, 1, 1)\} \subseteq 2^3$ is represented by the one-parameter word $f = (\lambda_0, 1, \lambda_0)$.

In general, an m-parameter set (or *combinatorial m-space*) $M \subseteq A^n$ is given by an m-parameter word $f \in (A \cup \{\lambda_0, \ldots, \lambda_{m-1}\})^n$. We require that each parameter λ_i, $i < m$, occurs at least once in f. The m distinct parameters $\lambda_0, \ldots, \lambda_{m-1}$ represent m mutually disjoint sets of moving coordinates. In order to avoid ambiguities we assume that $A \cap \{\lambda_i \mid i < m\} = \emptyset$, i.e., the set of constants $a \in A$ should be distinguished from the set of parameters $\lambda_i, i < m$. If $f \in (A \cup \{\lambda_0, \ldots, \lambda_{m-1}\})^n$ is an m-parameter word in A^n and $g \in (A \cup \{\lambda_0, \ldots, \lambda_{k-1}\})^m$ is a k-parameter word in A^m, the composition $f \cdot g \in (A \cup \{\lambda_0, \ldots, \lambda_{k-1}\})^n$ is the

k-parameter word in A^n resulting from replacing the parameter λ_i in f by g_i, the i-th component of g. In particular, for $k = 0$,

$$M = \{f \cdot (a_0, \ldots, a_{m-1}) \mid (a_0, \ldots, a_{m-1}) \in A^m\} \subseteq A^n$$

is the m-parameter set represented by f.

Clearly, two parameter words yield the same parameter set if they differ only by a permutation of their parameters. We get a rigid representation, i.e., a one-to-one correspondence between parameter sets and words, requiring the first occurrences of different parameters to be in increasing order, first λ_0, then λ_1, etc.

We summarize these ideas in a formal definition. The concept of parameter sets is due to Graham and Rothschild (1971), the formal calculus of parameter words has been introduced by Leeb (1973, unpublished).

Definition 3.1. For nonnegative integers $m \leq n$ we denote by $[A]\binom{n}{m}$ the set of all words (mappings) $f : n \to A \cup \{\lambda_0, \ldots, \lambda_{m-1}\}$ such that for every $j < m$ there exists $i < n$ with $f(i) = \lambda_j$, and $\min f^{-1}(\lambda_i) < \min f^{-1}(\lambda_j)$ for all $i < j < m$. The elements of $[A]\binom{n}{m}$ are called m-parameter words of length n over A. For $f \in [A]\binom{n}{m}$ and $g \in [A]\binom{m}{k}$ the composition $f \cdot g \in [A]\binom{n}{k}$ is defined by

$$(f \cdot g)(i) = \begin{cases} f(i) & \text{if} \quad f(i) \in A, \text{ and} \\ g(j) & \text{if} \quad f(i) = \lambda_j. \end{cases}$$

For $f \in [A]\binom{n}{m}$ the set

$$M = \{f \cdot g \mid g \in [A]\binom{m}{0}\} = f \cdot [A]\binom{m}{0}$$

is the m-parameter subset of A^n described by f. Observe that $[A]\binom{n}{0} = A^n$.

From the presentation of parameter sets via parameter words it easily follows that there are

$$S_m^n(t) = \frac{1}{m!} \sum_{i=0}^{m} (-1)^{m-i} \binom{m}{i} (t + i)^n$$

m-parameter subsets of A^n, putting again $|A| = t$. The numbers $S_m^n(t)$ are known as noncentral Stirling numbers of the second kind. Compare, e.g. Carlitz (1980) or Benzait and Voigt (1989) for discussion and combinatorial interpretation of these numbers.

Note that we have defined parameter words with respect to arbitrary finite alphabets (sets), including the empty and the one-element alphabet. Corresponding to different alphabets, parameter words admit different interpretations.

3.1.1 Parameter Words Over the Empty Alphabet

Partition lattices. Parameter words $f \in [\emptyset]\binom{n}{k}$ represent surjections from n onto k which are rigid in the sense that min $f^{-1}(\lambda_i) <$ min $f^{-1}(\lambda_j)$ for $i < j$. So they represent uniquely the equivalence relations on $n = \{0, \ldots, n-1\}$ with precisely k equivalence classes, and vice versa. The ith equivalence class is given by $f^{-1}(\lambda_i)$. Hence, $[\emptyset](n) = \bigcup_{k \leq n} [\emptyset]\binom{n}{k}$ is the set of all equivalence relations on n.

For $f \in [\emptyset]\binom{n}{m}$ and $g \in [\emptyset]\binom{n}{k}$ put $f \leq g$ if and only if there exists $h \in [\emptyset]\binom{m}{k}$ such that $g = f \cdot h$. Then $\Pi(n) \approx ([\emptyset](n), \leq)$ becomes the partition lattice of rank n, i.e., the lattice of equivalence relations on n.

3.1.2 Parameter Words Over a One-Element Alphabet

Sets. Consider $\phi : [\{0\}]\binom{n}{m} \to [n]^m$ defined by $\phi(f) = \{\min f^{-1}(\lambda_i) \mid i < m\}$. Obviously, ϕ is surjective. Assume $[n]^m$ to be given as the set of strictly ascending injections from m into n, i.e., $\phi(f) : m \to n$, where $\phi(f)(i) = \min f^{-1}(\lambda_i)$. Then ϕ has the property that for $f \in [\{0\}]\binom{n}{m}$ and $g \in [\{0\}]\binom{m}{k}$ it follows that $\phi(f \cdot g) = \phi(f) \cdot \phi(g)$. In the language of categories this is to say that ϕ is a functor.

Δ-systems. Parameter words $f \in [\{0\}]\binom{n}{m}$ represent families of m nonempty and disjoint subsets of $n = \{0, \ldots, n-1\}$, viz., $f^{-1}(\lambda_i)$, $i < m$. Then $f \cdot [\{0\}]\binom{m}{1}$ is the set of all nonempty unions of these m sets. Using the language of extremal problems, $[\{0\}]\binom{m}{k}$ is the set of strong Δ-systems with m terms.

3.1.3 Parameter Words Over a Two-Element Alphabet

Boolean lattices. Let $A = 2 = \{0, 1\}$. Every $f \in [2]\binom{n}{0}$ can be interpreted as the characteristic function of a subset of $n = \{0, \ldots, n-1\}$, where the letter 1 indicates the occurrence of an element in this subset. The inclusion of subsets imposes a lattice structure $\leq$ on $[2]\binom{n}{0}$. Provided with this order $([2]\binom{n}{0}, \leq)$ is isomorphic to the Boolean lattice $\mathcal{B}(n)$ of rank n. Parameter words $f \in [2]\binom{n}{k}$ represent $\mathcal{B}(k)$-sublattices in $\mathcal{B}(n)$, and vice versa. The composition $f \cdot g$ corresponds to taking a sublattice inside a sublattice. In Fig. 3.2 the images of $\mathcal{B}(1)$ under g and $f \cdot g$ are drawn boldfaced.

A partial order on Boolean sublattices of $\mathcal{B}(n)$ can be defined using the composition of parameter words. For $f \in [2]\binom{n}{k}$ and $g \in [2]\binom{n}{m}$ put $f \leq g$ if there exists $h \in [2]\binom{m}{k}$ such that $f = g \cdot h$.

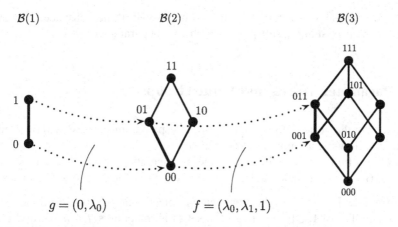

Fig. 3.2 The images of $\mathcal{B}(1)$ under $g = (0, \lambda_0)$ and $f \cdot g = (0, \lambda_0, 1)$ are drawn *boldfaced*. The $\mathcal{B}(2)$ and $\mathcal{B}(3)$-lattice are given by their Hasse-diagrams

3.1.4 Parameter Words Over a k-Element Alphabet

Arithmetic progressions. Let $A = k = \{0, \dots, k-1\}$ and consider the mapping $\Psi : A^n \to \mathbb{N}$ given by

$$\Psi(a_0, \dots, a_{n-1}) = \sum_{i < n} a_i.$$

Although Ψ is not one-to-one, we have that for every combinatorial line $f \in [k]\binom{n}{1}$ the set $\{\Psi(f \cdot i) \mid i < k\}$ is an arithmetic progression of length k.

If we choose the k-adic expansion of integers instead, i.e., $\kappa : A^n \to \mathbb{N}$ given by

$$\kappa(a_0, \dots, a_{n-1}) = \sum_{i < n} a_i k^i,$$

then

$$\{\kappa(f \cdot i) \mid i < k\}$$

is also an arithmetic progression for every combinatorial line f and, moreover, κ is one-to-one. In fact, κ is a bijection between A^n and the first k^n nonnegative integers.

3.1.5 Parameter Words Over GF(q)

Affine spaces. Let $GF(q)$, q a prime power, be the Gallois field with q elements. Then every m-parameter word $f \in [GF(q)]\binom{n}{m}$, resp., the corresponding

m-parameter subset in $GF(q)^n$, is an m-dimensional affine subspace. However, in general there exist affine subspaces which are not parameter subsets.

3.2   Parameter Words and Finite Groups

In this section the notion of parameter words (resp., parameter sets) will be slightly generalized allowing a finite group G to act on A.

Let G be a finite group, with unit element e, operating on A, i.e., there exists an operation $G \times A \to A$ such that $(\alpha \cdot \beta) \cdot a = \alpha \cdot (\beta \cdot a)$ for all $\alpha, \beta \in G$ and $a \in A$.

Definition 3.2. Let $m \leq n$ be nonnegative integers and G be any finite group acting on A. Then $[A, G]\binom{n}{m}$ denotes the set of all mappings $f : n \to A \cup (G \times \{\lambda_0, \ldots, \lambda_{m-1}\})$ such that

$$\begin{aligned}
f^{-1}(G \times \{\lambda_i\}) &\neq \emptyset & &\text{for every } i < m, \\
f(\min f^{-1}(G \times \{\lambda_i\})) &= (e, \lambda_i) & &\text{for every } i < m, \text{ and} \\
\min f^{-1}(G \times \{\lambda_i\}) &< \min f^{-1}(G \times \{\lambda_j\}) & &\text{for all } i < j < m.
\end{aligned}$$

The elements of $[A, G]\binom{n}{m}$ are *m-parameter words of length n over* $[A, G]$.

For $f \in [A, G]\binom{n}{m}$ and $g \in [A, G]\binom{m}{k}$ the composition $f \cdot g \in [A, G]\binom{n}{k}$ is defined by

$$(f \cdot g)(i) = \begin{cases} f(i) & \text{if } f(i) \in A \\ \alpha \cdot a & \text{if } f(i) = (\alpha, \lambda_j) \text{ and } g(j) = a \in A \\ (\alpha \cdot \beta, \lambda_\ell) & \text{if } f(i) = (\alpha, \lambda_j) \text{ and } g(j) = (\beta, \lambda_\ell). \end{cases}$$

What has changed is that parameters λ_i are labeled by group elements. To make these parameter words rigid the first occurrence of λ_i is labeled with the unit element e. The composition then is defined via group multiplication, resp., via the group action on A.

In fact, this is the original concept of parameter sets as it was introduced in Graham and Rothschild (1971).

3.2.1   *Parameter Words Over* $[\{0\}, GF(q)^*]$

Linear spaces.   Consider the multiplicative group $GF(q)^*$ operating on $\{0\}$, where 0 is the zero element of the Galois field $GF(q)$. Every $f \in [\{0\}, GF(q)^*]\binom{n}{m}$ represents an m-dimensional (homogeneous) linear subspace of the n-dimensional vector space over $GF(q)$. In general, there exist additional m-dimensional linear subspaces, except for $m = 1$, where we have bijective correspondence.

3.2.2   *Parameter Words Over* [{a}, G]

Dowling lattices. Using a different terminology, Dowling (1973) investigates parameter words $f \in [\{a\}, G]\binom{n}{m}$. The finite group G operates trivially on the singleton set $\{a\}$, where the a acts as a kind of annihilator. Put $[\{a\}, G](n) = \bigcup_{k<n}[\{a\}, G]\binom{n}{mk}$. For $f \in [\{a\}, G]\binom{n}{m}$ and $g \in [\{a\}, G]\binom{n}{k}$ put $f \leq g$ if and only if there exists $h \in [\{a\}, G]\binom{m}{k}$ such that $g = f \cdot h$. For the trivial group $G = \{e\}$ one easily observes that $([\{a\}, \{e\}](n), \leq)$ is isomorphic to $\Pi(n + 1)$, the partition lattice of rank $n + 1$. Dowling shows that, in general, $([\{a\}, G](n), \leq)$ is a geometric lattice of rank $n + 1$. Also, nonisomorphic groups yield nonisomorphic geometric lattices. Dowling also considers the problem to what extent $([\{a\}, G](n), \leq)$ is representable over a field K. He shows that this is the case if and only if G is isomorphic to a subgroup of the multiplicative group of K (necessity requires $n \geq 3$). The reader should compare Dowling's result with the example $[\{0\}, GF(q)^*]$.

3.2.3   *Parameter Words Over* [k, {e, π}]

Arithmetic progressions (revisited). Let $k > 0$. By $\pi : k \to k$ we denote the permutation given by $\pi(i) = k - 1 - i$. Consider the mapping $\Psi^* : k^n \to \mathbb{N}$ defined as

$$\Psi^*(a_0, \ldots, a_{n-1}) = \sum_{i<n} a_i 10^{k(n-1-i)}.$$

Obviously, Ψ^* is one-to-one, but not a bijection. However, here we have a bijective correspondence between the arithmetic progressions of length k in $\Psi^*(k^n)$ and the one-parameter words in $[k, \{e, \pi\}]\binom{n}{1}$.

Chapter 4
Hales-Jewett's Theorem

> *The streets of eighteenth-century England resounded with the*
> *voices of children chanting this simple rhyme:*
> *Tit, tat, toe, my first go,*
> *Three jolly butcher boys all in a row.*
> *Stick one up, stick one down,*
> *Stick one in the old man's crown.*
>   *This rhyme was recited by the winner of Noughts and*
> *Crosses, or Tic-Tac-Toe.*

> (from D. Olivatro (1984))

Tic-Tac-Toe is a game played by two people writing the symbols O and X in turn on a pattern of nine squares with the purpose of getting three such marks in a row. Of course, the traditional 3×3 Tic-Tac-Toe need not to have a winner, the second player can achieve a tie. But this does not remain true in general if we consider certain generalizations of the 3×3 Tic-Tac-Toe game. The t^n-game is played on a $t \times \ldots \times t$ (n times) array of points in n space, say on t^n. The rules are that each player in turn claims as his own a previously unclaimed element of t^n. He draws either a nought or a cross at this particular place. The game proceeds either until one of the players has claimed a complete line in t^n, in which case he wins, or until every element in t^n has been claimed, but no one has yet won, in which case the game is a tie.

Thereby a line forming a possible winning set is a subset $L \subseteq t^n, L = \{a_i \mid i < t\}$, where $a_i = (a_{i,0}, \ldots, a_{i,n-1})$, and for each $i < t$ either $a_{i,j} = b_i \in t$ for all $j < n$ or $a_{i,j} = j$ for all $j < n$ or $a_{i,j} = t - 1 - j$ for all $j < n$. Thus the winning sets are exactly the one-parameter words of length n over $[t, \{e, \pi\}]$, where $\pi : t \to t$ is given by $\pi(j) = t - 1 - j$.

Analyzing this game of Tic-Tac-Toe, A.W. Hales and R.I. Jewett (1963) proved a partition theorem for zero-parameter words, basically asserting that the first player always has a winning strategy, provided that n is sufficiently large with respect to t. This result will be proved in this chapter along with a brief discussion of bounds on

H.J. Prömel, *Ramsey Theory for Discrete Structures*,
DOI 10.1007/978-3-319-01315-2_4,
© Springer International Publishing Switzerland 2013

n and t, which enable us to draw some conclusion about the existence of winning and tying strategies.

But the influence of Hales-Jewett's theorem goes much beyond the analysis of Tic-Tac-Toe. In this chapter we will only give a glimpse on its consequences deriving some quite direct applications from this pigeon hole principle for parameter words, for example reproving van der Waerden's theorem on arithmetic progressions. But throughout the next chapters we shall meet several generalizations and ramifications of the Hales-Jewett theorem, and applications thereof, in various branches of Ramsey theory.

4.1 Hales-Jewett's Theorem

Throughout this section A denotes a fixed finite alphabet (set).

Convention. Let $f \in [A]\binom{m}{k}$ and $g \in [A]\binom{n}{\ell}$. Then $f \times g \in [A]\binom{m+n}{k+\ell}$ denotes the 'concatenation' of f and g, i.e.,

$$(f \times g)(i) = \begin{cases} f(i) & \text{if } i < m, \\ g(i-m) & \text{if } m \le i < n+m \text{ and } g(i-m) \in A, \text{ and} \\ \lambda_{k+j} & \text{if } m \le i < n+m \text{ and } g(i-m) = \lambda_j. \end{cases}$$

The theorem of Hales and Jewett (1963) is concerned with partitions of zero-parameter words, i.e., with partitions of A^n. We separate the special case of the two element alphabet, first considering partitions of 2^n only. On the one hand this will be done because this case is of particular interest via its interpretation as Boolean lattices, cf. Sect. 3.1.3, on the other hand because its proof is easier and will hopefully make some ideas more accessible.

Proposition 4.1. *Let m and r be positive integers. Then there exists a least positive integer $n = HJ(2, m, r)$ such that for every coloring $\Delta : [2]\binom{n}{0} \to r$ there exists a monochromatic m-parameter word $f \in [2]\binom{n}{m}$, which is to say that*

$$\Delta(f \cdot g) = \Delta(f \cdot h) \quad \text{for all } g, h \in [2]\binom{m}{0}.$$

Proof. The proof proceeds by induction on m. Let $m = 1$, r be an arbitrary positive integer and R be the following set of $r + 1$ many words each of length r:

$$R = \{ \quad (\quad 0, \quad 0, \quad \dots, \quad 0, \quad 0 \quad)$$
$$(\quad 0, \quad 0, \quad \dots, \quad 0, \quad 1 \quad)$$
$$\dots$$
$$(\quad 0, \quad 1, \quad \dots, \quad 1, \quad 1 \quad)$$
$$(\quad 1, \quad 1, \quad \dots, \quad 1, \quad 1 \quad) \quad \}.$$

For every r-coloring Δ of R there exist two words having the same color. Say

$$(0,\ldots,0,0,\ldots,0,1,\ldots,1)$$

and

$$(0,\ldots,0,1,\ldots,1,1,...,1).$$

Then the one-parameter word

$$(0,\ldots,0,\lambda_0,\ldots,\lambda_0,1,\ldots,1)$$

is monochromatic with respect to Δ.

Now assume that the assertion is true for some $m > 0$ and every r and choose

$$M = HJ(2,1,r) \quad \text{and} \quad N = HJ(2,m,r^{2^M})$$

and consider words of length $N + M$.

Let $\Delta : [2]\binom{M+N}{0} \to r$ be a coloring. This induces a coloring $\Delta_N : [2]\binom{N}{0} \to r^{2^M}$ on the tails of length N by coloring each tail with the sequence of colors it gets by varying over all possible initial pieces, i.e.,

$$\Delta_N(h) = \langle \Delta(g \times h) \mid g \in [2]\binom{M}{0}\rangle.$$

By choice of N there exists an m-parameter word $f_N \in [2]\binom{N}{m}$ which is monochromatic with respect to Δ_N. This means, fixing one initial piece, all insertions in f_N get the same color with respect to Δ.

Next consider $\Delta_M : [2]\binom{M}{0} \to r$ given by $\Delta_M(g) = \Delta(g \times (f_N \cdot h))$ for some (and hence for all) $h \in [2]\binom{m}{0}$. By the inductive assumption we know that there exists $f_M \in [2]\binom{M}{1}$ which is monochromatic with respect to Δ_M. Now the construction yields immediately that $f_M \times f_N \in [2]\binom{M+N}{1+m}$ is the desired monochromatic $(m+1)$-parameter word. $\qquad\square$

In the language of Boolean lattices, Proposition 4.1 says that for every r-coloring of the points of $\mathcal{B}(n)$ there exists a $\mathcal{B}(m)$-sublattice of $\mathcal{B}(n)$ which is monochromatic, provided that n was chosen sufficiently large. This can be visualized as in Fig. 4.1.

Now we prove Hales-Jewett's theorem for general (finite) alphabets.

Fig. 4.1 Point partition
property of Boolean lattices

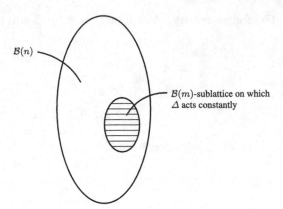

$\mathcal{B}(n)$

$\mathcal{B}(m)$-sublattice on which
Δ acts constantly

Theorem 4.2 (Hales, Jewett). *Let A be a finite alphabet and let m and r be positive integers. Then there exists a least positive integer $n = HJ(|A|, m, r)$ such that for every coloring $\Delta : [A]\binom{n}{0} \to r$ there exists an m-parameter word $f \in [A]\binom{n}{m}$, which is monochromatic.*

Proof. Let $t = |A|$. We show the following two inequalities:

(1) $HJ(t, m+1, r) \leq HJ(t, 1, r) + HJ(t, m, r^{t^{HJ(t,1,r)}})$
(2) $HJ(t+1, 1, r+1) \leq HJ(t, 1 + HJ(t+1, 1, r), r+1)$.

Together with the trivial observation that for every m and r we have that $HJ(1, m, r) = m$ (or using Proposition 4.1 instead) these two inequalities yield immediately the proof of Hales-Jewett's theorem by induction on t, m and r.

Proof of (1): We closely follow the approach from Proposition 4.1. Let $M = HJ(t, 1, r)$ and $N = HJ(t, m, r^{t^{HJ(t,1,r)}})$ and consider $\Delta : [A]\binom{M+N}{0} \to r$. This induces a coloring $\Delta_N : [A]\binom{N}{0} \to r^{t^M}$ by

$$\Delta_N(h) = \langle \Delta(g \times h) \mid g \in [A]\binom{M}{0}\rangle.$$

By choice of N there exists an m-parameter word $f_N \in [A]\binom{N}{m}$ which is monochromatic with respect to Δ_N. Next consider $\Delta_M : [A]\binom{M}{0} \to r$, given by

$$\Delta_M(g) = \Delta(g \times (f_N \cdot h)) \quad \text{for some (and hence all) } h \in [A]\binom{m}{0}.$$

By choice of M there exists $f_M \in [A]\binom{M}{1}$ which is monochromatic with respect to Δ_M. Now $f_M \times f_N \in [A]\binom{M+N}{m+1}$ proves that inequality (1) is valid.

Proof of (2): Let $N = HJ(t, 1 + HJ(t+1, 1, r), r+1)$, $b \notin A$ and consider $\Delta : [A \cup \{b\}]\binom{N}{0} \to r+1$. Let $\Delta_A = \Delta\lceil[A]\binom{N}{0}$. By choice of N there exists $f_A \in [A]\binom{N}{1+M}$, where $M = HJ(t+1, 1, r)$, which is monochromatic with respect to Δ_A. Say, $\Delta_A\lceil f_A \cdot [A]\binom{1+M}{0} \equiv r$. If $\Delta(f_A \cdot (b \times g)) = r$ for some $g \in [A \cup \{b\}]\binom{M}{0}$, then replace all b's in $f_A \cdot (b \times g)$ by λ_0 and call the resulting one-parameter word f. Clearly, $f \in [A \cup \{b\}]\binom{N}{1}$ and $\Delta] f \cdot [A \cup \{b\}]\binom{1}{0}$ is constant. If no such $g \in [A \cup \{b\}]\binom{N}{0}$ exists, consider $\Delta_M : [A \cup \{b\}]\binom{M}{0} \to r$ defined by $\Delta_M(g) = \Delta(f_A \cdot (b \times g))$. By choice of M there exists $f_M \in [A \cup \{b\}]\binom{M}{1}$ monochromatic with respect to Δ_M. In this case $f_A \cdot (b \times f_M)$ proves that inequality (2) is valid. $\square$

The inequalities (1) and (2) immediately show that the bound on the function $n = HJ(|A|, m, r)$ which we get from this proof of Hales-Jewett's theorem is not primitive recursive. Whether this reflects the truth or whether this is just a consequence of the double induction used in the proof was an open problem for quite some time, until Shelah (1988) in a celebrated paper came up with a different proof of Hales-Jewett's theorem which implied that the function $n = HJ(|A|, m, r)$ is primitive recursive.

4.2   Some Applications

4.2.1   Arithmetic Progressions

In some sense, Hales-Jewett's theorem reveals the combinatorial heart of van der Waerden's theorem on arithmetic progressions, stripping the arithmetic structure of the problem. Consider the alphabet $A = t = \{0, \ldots, t - 1\}$. The mapping $\Psi : A^n \to n(t - 1)$ with $\Psi(a_0, \ldots, a_{n-1}) = \sum a_i$ has the property that it maps every one-parameter word onto a t-term arithmetic progression (cf. Sect. 3.1.4). Hence, Hales-Jewett's theorem implies immediately van der Waerden's theorem on arithmetic progressions:

Theorem 4.3 (van der Waerden). *Let r and t be positive integers. Then there exists a least positive integer $n = W(t, r)$ such that for every coloring Δ : $[1, n] \to r$ there exists a monochromatic t-term arithmetic progression.* $\square$

4.2.2   Gallai-Witt's Theorem

A multidimensional version of van der Waerden's theorem was proved independently by Gallai (=Grünwald), cf. Rado (1943), and Witt (1952).

Let $X = \{x_0, \ldots, x_{t-1}\} \subseteq \mathbb{R}^m$ be a finite set of points in the Euclidean m-space. A homothetic mapping (homothety) is a mapping $h : \mathbb{R}^m \to \mathbb{R}^m$ of the form $h(x) = a + dx$, where $a \in \mathbb{R}^m$ is the translation vector and $d \in \mathbb{R} \setminus \{0\}$ describes a dilatation. The image $h(X) \subseteq \mathbb{R}^m$ is a homothetic copy of X.

Theorem 4.4 (Gallai, Witt). *Let r, m be positive integers and $X \subseteq \mathbb{R}^m$ be a finite set. Then there exists a finite set $Y \subseteq \mathbb{R}^m$ such that for every coloring $\Delta : Y \to r$ there exists a homothetic copy of X in Y which is monochromatic.*

Proof. Here the same idea applies as in proving van der Waerden's theorem. Put $A = X$ and let $n = HJ(|A|, 1, r)$. Consider $\Psi : A^n \to \mathbb{R}^m$ given by $\Psi(a_0, \ldots, a_{n-1}) = \sum_{i<n} a_i$ and let $Y = \Psi(A^n)$.

Now let $\Delta : Y \to r$ be a coloring. This induces a coloring $\Delta^* : A^n \to r$ via $\Delta^*(a_0, \ldots, a_{n-1}) = \Delta(\sum_{i<n} a_i)$. By choice of n there exists $f \in [A]\binom{n}{1}$ being monochromatic with respect to Δ^*. Put $a = \sum\{f(i) \mid f(i) \neq \lambda_0\}$ and $d = |\{i \mid f(i) = \lambda_0\}|$. Then, obviously $\Delta\rceil\{a + dx \mid x \in X\}$ is constant.   □

4.2.3 Deuber's (m, p, c)-Sets

The next application of Hales-Jewett's theorem extends the Gallai-Witt theorem and completes the proof of Rado's Theorem 2.8.

Let m, p, c be positive integers. Recall from Sect. 2.5 that a set $M \subseteq N$ is an (m, p, c)-set if there exist positive integers $x_0, \ldots, x_m$ such that

$$M = M_{p,c}(x_0, \ldots, x_m)$$

$$= \{cx_i + \sum_{j=i+1}^{m} \xi_j x_j \mid \xi_j \in [-p, p] \cap \mathbb{Z} \text{ for every } j \in [i+1, m] \text{ and } i \leq m\}.$$

Helpful for our purposes is to visualize an (m, p, c)-set in the following way:

$$cx_0 + \xi_1 x_1 + \xi_2 x_2 + \ldots + \xi_m x_m$$
$$cx_1 + \xi_2 x_2 + \ldots + \xi_m x_m$$
$$cx_2 + \ldots + \xi_m x_m$$
$$\vdots$$
$$cx_m$$

where $\xi_j \in [-p, p] \cap \mathbb{Z}$ for $j \in [1, m]$.

We will sometimes refer to this figure speaking, e.g., of the kth row of an (m, p, c)-set, which is the row that starts with cx_k, i.e., we start with a 0th row. Observe that besides the leading coefficient c each row is a multiple arithmetic progression.

We now use Hales-Jewett's theorem to prove the partition theorem for (m, p, c)-sets.

Theorem 4.5 (Deuber). *Let m, p, c and r be positive integers. Then there exist positive integers n, q and d such that for every coloring $\Delta : \mathbb{N} \to r$ of the positive integers every (n, q, d)-set $N \subseteq \mathbb{N}$ contains a monochromatic (m, p, c)-set.*

Proof. First we show:

(1) Let m, p, c, r and $k \leq m$ be positive integers. Then there exist positive integers n, q, and d with the following property:
Let N be an (n, q, d)-set. Then for every coloring $\Delta : N \to r$ there exists an (m, p, c)-set $M \subseteq N$ such that on each of the first k rows of M the coloring Δ is constant, i.e., $\Delta(x) = \Delta(y)$ whenever x, y are elements of the ith row of M for some $i \leq k$.

We prove (1) by induction on k. First consider the case $k = 0$. Let $q = cp$, $d = c^2$, $A = [-p, p]$ and let $n = HJ(|A|, m, r)$ be according to Hales-Jewett's theorem. Let $N = N_{q,d}(y_0, \dots, y_n)$ be an (n, q, d)-set and $\Delta : N \to r$ an r-coloring of N. We define a coloring $\Delta' : [A]\binom{n}{0} \to r$ by

$$\Delta'(\xi_1, \dots, \xi_n) = \Delta(dy_0 + c \sum_{l=1}^{n} \xi_i y_i).$$

Observe that the definition of an (n, q, d)-set, together with choice of $q = cp$, implies that the sums on the right hand side are indeed contained in N. By choice of n there exists an $f \in [A]\binom{n}{m}$ which is monochromatic with respect to Δ'. Now consider the (m, p, c)-set M defined by $M = M_{p,c}(z_0, z_1, \dots, z_m)$, where

$$z_0 = cy_0 + \sum_{i : f(i) \in A} f(i) \, y_{1+i},$$

and

$$z_{1+j} = c \sum_{i : f(i) = \lambda_j} y_{1+i} \quad \text{for } j < m.$$

Then the fact that $f \in [A]\binom{n}{m}$ is monochromatic with respect to Δ' implies that Δ is constant on each of the 0th rows of M.

Now assume that (1) is valid for some $k \geq 0$. We proceed similarly as in the case $k = 0$. Let $q = cp^2$, $d = c^2$, $A = [-p, p]$ and let $n = HJ(|A|, m - k, r) + k$ be according to Hales-Jewett's theorem. We apply the induction assumption for k and with respect to $m \leftarrow n$, $p \leftarrow q$, and $c \leftarrow d$ in order to see that by starting with appropriate parameters n', q', d' we may assume that every (n', q', d')-set N' and coloring $\Delta : N' \to r$ contains an (n, q, d)-set N such that Δ is constant on each of the first k rows of N. To handle the $(k + 1)$st row define a coloring $\Delta' : [A]\binom{n-k}{0} \to r$ by

$$\Delta'(\xi_{k+1}, \ldots, \xi_n) = \Delta(dy_k + c \sum_{i=k+1}^{n} \xi_i y_i).$$

By choice of n there exists an $f \in [A]\binom{n-k}{m-k}$ which is monochromatic with respect to Δ'. We define an (m, p, c)-set by $M = M_{p,c}(cy_0, \ldots, cy_{k-1}, z_k, \ldots, z_m)$, where

$$z_k = cy_k + \sum_{f(i) \in A} f(i) \, y_{k+1+i}, \text{ and}$$

$$z_{k+1+j} = c \sum_{f(i) = \lambda_j} y_{k+1+i} \quad \text{for } j < m - k.$$

Observe that for $i \leq k$ the ith row of M is a subset of the ith row of N. (To see this use that $q = cp^2$.) Hence, Δ is monochromatic on these rows. For row $k + 1$, on the other hand, the fact that $f \in [A]\binom{n}{m}$ is monochromatic with respect to Δ' implies that Δ is constant on the $(k + 1)$st row of M, completing the proof of (1).

To complete the proof of the theorem, put $\tilde{m} = rm$ and use (1) in order to observe that there exist n, q, and d such that every (n, q, d)-set N contains for every r-coloring $\Delta : N \to r$ an $(\tilde{m}, p, c)$-set $\tilde{M} = \tilde{M}_{p,c}(x_0, \ldots, x_{\tilde{m}})$ so that Δ is constant on each row of $\tilde{M}$. By the pigeon hole principle, then, there exist $m + 1$ rows, say $i_0 < \ldots < i_m$ on which Δ has the same color. Hence, the (m, p, c)-set $M = M_{p,c}(x_{i_0}, \ldots, x_{i_m}) \subseteq N$ is monochromatic with respect to Δ. $\square$

4.2.4 Idempotents in Finite Algebras

Let a be a nonnegative integer and let $\alpha = (\alpha_0, \ldots, \alpha_a)$ be a sequence of positive integers. An algebra of type α is a pair $(B, \mathcal{B})$, where B is a nonempty set and $\mathcal{B} : B^{\alpha_i} \to B$, for $i \leq a$, is an α_i-ary operation (by abuse of language we use the same $\mathcal{B}$ for all i). An algebra $(A, \mathcal{A})$ of type α is a subalgebra of $(B, \mathcal{B})$ if $A \subseteq B$ and A is closed under the operations $\mathcal{B}$.

Theorem 4.6. *Let $\mathcal{K}$ be a class of finite algebras of type α which is closed under finite products and such that every member $(A, \mathcal{A})$ of $\mathcal{K}$ contains idempotents only, i.e., $A(x, \ldots, x) = x$ for every $x \in A$. Let r be a positive integer and $(A, \mathcal{A}) \in \mathcal{K}$. Then there exists a $(B, \mathcal{B}) \in \mathcal{K}$ such that for every coloring $\Delta : B \to r$ there exists a monochromatic subalgebra of $(B, \mathcal{B})$ which is isomorphic to $(A, \mathcal{A})$.*

Proof. Let $n = HJ(|A|, 1, r)$ and choose $(B, \mathcal{B}) = (A, \mathcal{A})^n$. Recall that $\mathcal{K}$ is closed under finite products. Hence, $(B, \mathcal{B}) \in \mathcal{K}$. Moreover, by Hales-Jewett's theorem we know that $(B, \mathcal{B})$ has the desired property. $\square$

We will abbreviate this result by saying that the class $\mathcal{K}$ has the *partition property with respect to points*. Theorem 4.6 occurs in Ježek and Nešetřil (1983) and Prömel and Voigt (1981b).

4.2.5   Lattices and Posets

Theorem 4.6 applies in particular to a variety of finite lattices. Some of them we will mention explicitly. For basic facts about lattices we refer the reader to Birkhoff (1967) or Grätzer (1998).

Distributive lattices. Although the partition property of points in distributive lattices follows from Theorem 4.6, it can already be derived from Proposition 4.1 using that distributive lattices are exactly the sublattices of Boolean lattices. Distributive lattices will be discussed in more detail in Sect. 5.2.3.

Partially ordered sets (posets). It can easily be seen that every poset can be embedded (as an order) in some Boolean lattice. So we get from Proposition 4.1 that the class of all finite posets has the partition property with respect to points. In full length:

Let r be a positive integer and Q be a finite poset. Then there exists a finite poset P such that for every coloring $\Delta : P \to r$ of the points of P there exists a Q-subposet of P which is monochromatic.

A slight generalization of Theorem 4.6 covering also relational systems of a certain type and in particular covering posets, is given in Pouzet and Rosenberg (1985).

Partition lattices. By a celebrated theorem of Pudlák and Tůma (1980) every finite lattice can be embedded into some partition lattice $\Pi(n)$. Using that the class of all finite lattices has the partition property with respect to points we can derive immediately from this that the class of all finite partition lattices has also the partition property with respect to points, i.e., for every pair m and r of positive integers there exists a positive integer $n = n(m, r)$ such that for every coloring $\Delta : \Pi(n) \to r$ of the points of $\Pi(n)$ with r colors there exists a $\Pi(m)$-sublattice of $\Pi(n)$ which is monochromatic. This situation is depicted in Fig. 4.2.

4.3   A ∗-Version

In this section not only colorings of zero-parameter words of one fixed length are considered, as in Hales-Jewett's theorem, but words of variable length (where a ∗ indicates the end of a word). Such ∗-parameter words were originally introduced by Voigt (1980) to prove a partition theorem for finite Abelian groups. They will also

Fig. 4.2 Point partition
property of partition lattices

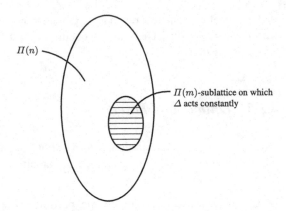

$\Pi(n)$

$\Pi(m)$-sublattice on which
Δ acts constantly

be a quite useful tool in proving a higher dimensional analogue to Hales-Jewett's
theorem (cf. Chap. 5).

Convention. Let $'*'$ be a symbol not contained in $A \cup \{\lambda_0, \ldots, \lambda_{m-1}\}$ and let
$[A]^* \binom{n}{m}$ denote the set of all m-parameter words f of length n over $A \cup \{*\}$
satisfying the condition

$$f(i) = * \quad \text{for some } i < n \text{ implies that} \quad f(j) = * \quad \text{for all } i \leq j < n.$$

Hence, $[A]^* \binom{n}{m}$ can be viewed as the set of m-parameter words of length at most n
over A. Note that in this sense $[A]\binom{n}{m} \subseteq [A]^* \binom{n}{m}$.

For $f \in [A]^* \binom{n}{m}$ and $g \in [A]^* \binom{m}{k}$ the composition $f \cdot g \in [A]^* \binom{n}{k}$ is defined by

$$(f \cdot g)(i) = \begin{cases} * & \text{if there exists } j < i \text{ such that } (f \cdot g)(j) = *, \\ f(i) & \text{if } f(i) \in A \cup \{*\} \text{ and } (f \cdot g)(j) \neq * \text{ for all } j < i, \\ g(j) & \text{if } f(i) = \lambda_j \text{ and } (f \cdot g)(j) \neq * \text{ for all } j < i \end{cases}$$

Intuitively, the composition $f \cdot g$ interpreted as the insertion of g into the parameters
of f is performed as long as possible, eventually $*$'s are filled in.

Theorem 4.7. *Let A be a finite alphabet and let m, r be positive integers. Then
there exists a positive integer $n = HJ^*(|A|, m, r)$ such that for every coloring Δ :
$[A]^* \binom{n}{0} \to r$ there exists a monochromatic $f \in [A]^* \binom{n}{m}$, i.e., $\Delta(f \cdot g) = \Delta(f \cdot h)$
for all $g, h \in [A]^* \binom{m}{0}$.*

Proof. Let $n_{mr} = mr$ and $n_{mr-j} = HJ(|A|, n_{mr-j+1} - mr + j, r) + mr - j$. Choose
$n = n_0$ and let $\Delta : [A]^* \binom{n}{0} \to r$ be a coloring.

For $g \in [A]^* \binom{k}{0}$ let $*(g)$ denote the number of $*$'s at the end of g, i.e., $*(g) = k - 1 - \max\{i < k \mid g(i) \in A\}$ with $\max \emptyset = -1$. For every $i \leq k$ put

$$[A]^i \binom{k}{0} = \{g \in [A]^* \binom{k}{0} \mid *(g) = i\}.$$

In particular,

$$\bigcup_{i \leq k} [A]^i \binom{k}{0} = [A]^* \binom{k}{0}.$$

First we prove inductively that for every $j \leq mr$ there exists $f_j \in [A]\binom{n}{n_{j+1}}$ such that for every $g, h \in \bigcup_{i \leq j} [A]^i \binom{n_{j+1}}{0}$ satisfying $*(g) = *(h)$ we have $\Delta(f_j \cdot g) = \Delta(f_j \cdot h)$.

For $j = 0$, i.e., considering only words without $*$'s at the end, this is Hales-Jewett's theorem. So assume that the assertion is true for some $j < mr$ and let $\Delta^{j+1} : [A]^{j+1}\binom{n_{j+1}}{0} \to r$ be given by $\Delta^{j+1}(g) = \Delta(f_j \cdot g)$. By choice of $n_{j+1} = HJ(|A|, n_{j+2} - j - 1, r) + j + 1$ and Hales-Jewett's theorem there exists

$$f' \in [A]\binom{n_{j+1}-j-1}{n_{j+2}-j-1}$$

which is monochromatic with respect to Δ^{j+1}. Then, obviously, $f_{j+1} = f_j \cdot (f' \times (\lambda_{n_{j+2}-j-1}, \ldots, \lambda_{n_{j+2}-1}))$ fulfills the requirement of the induction.

Choosing $j = mr$ we get $f_{mr} \in [A]\binom{n}{mr}$ such that all $g, h \in [A]^*\binom{mr}{0}$ satisfying $*(g) = *(h)$ have the same color with respect to Δ. This defines an r-coloring Δ' of the integers $0, \ldots, mr$ by $\Delta'(i) = \Delta(f_{mr} \cdot g)$ for any g with $*(g) = i$. By the pigeonhole principle we get $0 \leq i_0 < \ldots < i_m \leq mr$ in one color. Now let $f'' \in [A]\binom{mr}{m}$ be given by $f(i) = a$ for some $a \in A$ if $i < i_0$, $f(i) = \lambda_j$ if $i_j \leq i < i_{j+1}$ and $f(i) = *$ for $i_m \leq i$. Clearly, $f = f_{mr} \cdot f''$ has the desired properties. $\square$

Chapter 5
Graham-Rothschild's Theorem

5.1 Graham-Rothschild's Theorem

The Graham and Rothschild theorem (1971) is concerned with partitions of k-parameter words in A^n. It generalizes Hales-Jewett's theorem to higher dimensions.

Theorem 5.1 (Graham, Rothschild). *Let A be a finite alphabet, let G be a finite group acting on A and let k, m and r be positive integers. Then there exists a positive integer $n = GR(|A|, |G|, k, m, r)$ such that for every coloring $\Delta : [A, G]\binom{n}{k} \to r$ there exists an $f \in [A, G]\binom{n}{m}$ which is monochromatic, i.e.,*

$$\Delta(f \cdot g) = \Delta(f \cdot h) \quad \text{for all } g, h \in [A, G]\binom{m}{k}.$$

Proof. Without loss of generality we may assume that $|A| \geq 1$. We proceed by induction on k. The case $k = 0$ is settled by Hales-Jewett's theorem. So we can assume that the theorem is valid for some $k - 1 \geq 0$, for every finite alphabet B with $|B| \geq 1$ and every number r' of colors.

We use the $*$-version of Hales-Jewett's theorem (Theorem 4.7). Let $x = HJ^*(|A|, m, r)$ and $n_x = x + k$. For $0 < j \leq x$ let

$$n_{x-j} = GR(|A| + |G|, |G|, k - 1, n_{x-j+1} - x + j - 1, r^{|A|^{(x-j)}}) + x - j + 1$$

which exists according to our inductive hypothesis. We claim that $n = n_0$ is as required in the theorem.

Let $\Delta : [A, G]\binom{n}{k} \to r$ be a coloring. For $g \in [A, G]\binom{n}{k}$ let $In(g) \in [A]^*\binom{n}{0}$ be given by $In(g)(i) = g(i)$ for $i < \min g^{-1}(e, \lambda_0)$ and $In(g)(i) = *$ otherwise. Moreover, let $|In(g)| = \min g^{-1}(e, \lambda_0)$.

H.J. Prömel, *Ramsey Theory for Discrete Structures*,
DOI 10.1007/978-3-319-01315-2_5,
© Springer International Publishing Switzerland 2013

First, we prove inductively that for every $j \leq x$ there exists $f_j \in [A, G]\binom{n}{n_j}$ such that for every $g, h \in [A, G]\binom{n_j}{k}$ with $In(g) = In(h)$ and $|In(g)| < j$ we have $\Delta(f_j \cdot g) = \Delta(f_j \cdot h)$.

For $j = 0$ the assertion holds vacuously. So assume that the claim is true for some $j < x$ and let

$$\Delta^j : [A \cup (G \times \{\lambda_0\}), G]\binom{n_j - j - 1}{k - 1} \to r^{|A|^j}$$

be given by

$$\Delta^j(g) = \langle \Delta(f_j \cdot g') \mid (g'(0), \dots, g'(j - 1)) \in A^j, g'(j) = (e, \lambda_0),$$
$$\text{and } g'(j + \ell) = g(\ell - 1) \text{ for } 1 \leq \ell \leq n_j - j - 1 \rangle,$$

where we assume that the parameters in $g \in [A \cup (G \times \{\lambda_0\}), G]\binom{n_j - j - 1}{k - 1}$ are numbered from 1 to k to make them disjoint from the alphabet. By choice of

$$n_j = GR(|A| + |G|, |G|, k - 1, n_{j+1} - j - 1, r^{|A|^j}) + j + 1$$

there exists

$$f' \in [A \cup (G \times \{\lambda_0\}), G]\binom{n_j - j - 1}{n_{j+1} - j - 1}$$

which is monochromatic with respect to Δ^j. Then let

$$f'' \in [A, G]\binom{n_j}{n_{j+1}}$$

be given by

$$f''(i) = \begin{cases} (e, \lambda_i) & \text{if } i \leq j, \\ f'(i - j - 1) & \text{if } j < i < n_j \text{ and } f'(i - j - 1) \in A, \text{ and} \\ (\alpha, \lambda_{j+k}) & \text{if } j < i < n_j \text{ and } f'(i - j - 1) = (\alpha, \lambda_k). \end{cases}$$

Then $f_{j+1} = f_j \cdot f''$ fulfills the requirement of the inductive step.

Eventually we obtain $f_x \in [A, G]\binom{n}{x+k}$ such that all $g, h \in [A, G]\binom{x+k}{k}$ satisfying $In(g) = In(h)$ are colored the same with respect to Δ. This defines a coloring $\Delta' : [A]^* \binom{x}{0} \to r$ by $\Delta'(g) = \Delta(f_x \cdot h)$ for some (and hence every) $h \in [A, G]\binom{x+k}{k}$ satisfying $In(h) = g$. By choice of $x = HJ^*(|A|, m, r)$ we get some $f^* \in [A]^* \binom{x}{m}$ which is monochromatic with respect to Δ'. Define $f^{**} \in [A, G]\binom{x+k}{m}$ by $f^{**}(i) = f^*(i)$ if $f^*(i) \in A$, $f^{**}(i) = (e, \lambda_j)$ if $f^*(i) = \lambda_j$, and $f^{**}(i) = a$ for some $a \in A$ otherwise.

Now let $f = f_x \cdot f^{**}$. Then $f \in [A, G]\binom{n}{m}$ has the desired properties.   □

Fig. 5.1 Ramsey's theorem

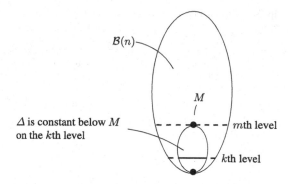

$\mathcal{B}(n)$

M

Δ is constant below M
on the kth level

mth level

kth level

Most applications of the Graham-Rothschild theorem consider the special case $G = \{e\}$, the trivial group. We formulate this as a corollary:

Corollary 5.2. *Let A be a finite alphabet and k, m and r be positive integers. Then there exists a positive integer $n = GR(|A|, k, m, r)$ such that for every coloring $\Delta :$ $[A]\binom{n}{k} \to r$ there exists an $f \in [A]\binom{n}{m}$ which is monochromatic.* □

5.2 Applications

5.2.1 Ramsey's Theorem

Let $A = 1 = \{0\}$ and consider the mapping $\phi : [1]\binom{n}{k} \to [n]^k$ defined by $\phi(g) = \{\min g^{-1}(\lambda_i) \mid i < k\}$, cf. Sect. 3.1.2. Clearly, ϕ is surjective.

Let $n = GR(1, k, m, r)$, for positive integers k, m and r, according to (the corollary to) Graham-Rothschild's theorem and let $\Delta : [n]^k \to r$ be a coloring of the k-subsets of n. Define $\Delta_\phi : [1]\binom{n}{k} \to r$ by $\Delta_\phi(g) = \Delta(\phi(g))$. By choice of n there exists $f_\phi \in [1]\binom{n}{m}$ which is monochromatic with respect to Δ_ϕ.

Observing that $\phi(f_\phi \cdot g) = \phi(f_\phi) \cdot \phi(g)$ for every $g \in [1]\binom{m}{k}$ (considering sets as rigid injections) this implies that $\Delta \upharpoonright [\phi(f_\phi)]^k$ is constant. Choosing $M = \phi(f_\phi)$ we have reproved:

Theorem 5.3 (Ramsey). *Let k, m and r be positive integers. Then there exists a least positive integer $n = RAM(k, m, r)$ such that for every coloring $\Delta : [n]^k \to \tau$ there exists $M \in [n]^m$ which is monochromatic.* □

Ramsey's theorem can be illustrated as in Fig. 5.1; the reader should also compare this with Fig. 4.1.

Fig. 5.2 The dual Ramsey
theorem

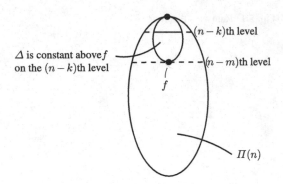

Δ is constant above f
on the $(n-k)$th level

$(n-k)$th level

$(n-m)$th level

f

$\Pi(n)$

5.2.2   The Dual Ramsey Theorem

Ramsey's theorem deals with subsets which can be interpreted as strictly ascending
(rigid) injections. The dual Ramsey theorem deals with rigid surjections, i.e.,
partitions. Consider $\Pi(n)$, the lattice of equivalence relations (partitions) on $n =$
$\{0,\ldots,n-1\}$. Recall that $\Pi(n) \approx ([\emptyset](n), \leq)$ where $f \leq g$, for some $f \in [\emptyset]\binom{n}{m}$
and some $g \in [\emptyset]\binom{n}{k}$, if there exists $h \in [\emptyset]\binom{m}{k}$ such that $g = f \cdot h$ (cf.
Sect. 3.1.1). In particular, $\Pi\binom{n}{k}$ (denoting the set of all those equivalence relations
on n which have precisely k classes) is represented by $[\emptyset]\binom{n}{k}$, and vice versa. Using
this interpretation we derive immediately from the Graham-Rothschild theorem a
dual Ramsey theorem. We first give a picture (Fig. 5.2). Here, the reader should
compare Fig. 5.1 with Fig. 4.2.

Theorem 5.4 (Dual Ramsey theorem).   *Let k, m and r be positive integers. Then
there exists a positive integer $n = DR(k,m,r)$ such that for every coloring Δ :
$\Pi\binom{n}{k} \to r$ there exists $f \in \Pi\binom{n}{m}$ which is monochromatic, i.e., all equivalence
relations on n with exactly k classes which are coarser than f are colored the
same.*   □

5.2.3   Distributive Lattices

The point partition property of distributive lattices follows already from
Hales-Jewett's theorem (cf. Sect. 4.2.5). Via the same arguments, i.e., using that
every finite distributive lattice can be embedded into some Boolean lattice $\mathcal{B}(n)$
and that Boolean lattices are represented by parameter words over the alphabet
$A = 2 = \{0, 1\}$, we get from the Graham-Rothschild theorem:

Theorem 5.5.   *Let $D = \mathcal{B}(k)$ be a Boolean lattice, let E be a distributive lattice an
d let r be a positive integer. Then there exists a distributive lattice F such that for*

every coloring $\Delta : \binom{F}{D} \to r$ of the D-sublattices of F there exists a E-sublattice $\tilde{E}$ in F such that all D-sublattices of $\tilde{E}$ are colored the same. □

This result leads to the question for which distributive lattices D, besides Boolean lattices, an analogous property is true. The answer is as easy to state as to prove: for no other distributive lattices.

To see this let D be a finite distributive lattice which is not a Boolean lattice and let $\mathcal{B}(D)$ be the smallest Boolean lattice in which D can be embedded, i.e., $\mathcal{B}(D) = \mathcal{B}(m)$ for some m. Observe that there exists a D-sublattice in $\mathcal{B}(D)$ in which the atom $\{0\}$ is contained and another D-sublattice in which the atom $\{0\}$ is not contained.

Now let $\mathcal{B}(n)$ be an arbitrary Boolean lattice and $\tilde{D}$ be a D-sublattice of $\mathcal{B}(n)$. Consider the $\mathcal{B}(D)$-sublattice $\mathcal{B}(\tilde{D})$ in $\mathcal{B}(n)$ which is generated by $\tilde{D}$. Say, $\mathcal{B}(\tilde{D})$ has the set $C \subseteq n$ as its minimum and mutually disjoint and nonempty sets $B_0, \ldots, B_{m-1}$ as atoms. Assume that B_0 is the lexicographic smallest of these atoms with respect to the natural order on n. Then color $\tilde{D}$ with color 1 if B_0 is contained in $\tilde{D}$, with color 0 else.

Continue along these lines. At the end, every $\mathcal{B}(D)$-sublattice of $\mathcal{B}(n)$ contains a 0-colored copy of D as well as a 1-colored copy. Summarizing this yields:

Observation 5.6. *Let D be a distributive lattice which is not a Boolean lattice. Then there exists a Boolean lattice $E = \mathcal{B}(m)$ such that for every distributive lattice F there exists a coloring $\Delta : \binom{F}{D} \to 2$ of the D-sublattices of F such that every E-sublattice of F contains two different colored D-sublattices.* □

5.2.4 Finite Unions and Finite Sums

The particular case $A = \{0\}$ and $k = 1$ of the Graham-Rothschild theorem can be stated as follows (cf. Sect. 3.1.2):

Theorem 5.7 (Finite union theorem). *Let m and r be positive integers. Then there exists a positive integer $n = FU(m, r)$ such that for every coloring $\Delta : \mathcal{B}(n) \to r$ there exist m mutually disjoint and nonempty subsets $B_0, \ldots, B_{m-1} \in \mathcal{B}(n)$ such that for all non empty $I, J \subseteq m$ it follows that*

$$\Delta\left(\bigcup_{i \in I} B_i\right) = \Delta\left(\bigcup_{j \in J} B_j\right).$$

□

Using a diagram this theorem can be presented as in Fig. 5.3. The reader should compare this diagram with Fig. 4.1. This makes clear that the finite union theorem is a projective analogue to the point partition theorem for Boolean lattices, viz. Proposition 4.1. Thus these two results correspond in the same way as the Rado-Folkman-Sanders theorem corresponds to Hilbert's cube lemma.

Fig. 5.3 The finite union
theorem

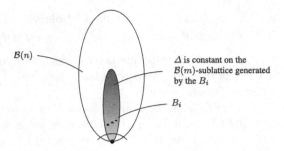

To deduce the Rado-Folkman-Sanders theorem from the finite union theorem
we consider the mapping $\kappa : \mathcal{B}(n) \to 2^n$ given by $\kappa(B) = \sum_{i \in B} 2^i$ for every
$B \subseteq n$. Obviously, κ is a bijection, in fact, κ^{-1} gives the binary expansion of positive
integers. Observing that for disjoint sets B_0 and B_1 we have $\kappa(B_0) + \kappa(B_1) = \kappa(B_0 \cup B_1)$ one gets:

Theorem 5.8 (Rado, Folkman, Sanders). *Let m and r be positive integers. Then
there exists a positive integer $n = FS(m, r)$ such that for every coloring $\Delta : n \to r$
there exist m (mutually distinct) positive integers $a_0, \ldots, a_{m-1}$ such that for all
nonempty $I, J \subseteq m$ it follows that*

$$\Delta(\sum_{i \in I} a_i) = \Delta(\sum_{j \in J} a_j).$$

$\square$

5.2.5 Linear and Affine Spaces

Recalling that every m-parameter word $f \in [GF(q)]\binom{n}{m}$ corresponds to an
m-dimensional affine subspace of $(GF(q))^n$ (cf. Sect. 3.1.5) and that there is a
one-to-one correspondence between the zero-parameter words of length n over
$GF(q)$ and the affine points in the affine space $(GF(q))^n$ we obtain already from
Hales-Jewett's theorem a partition theorem for points in affine spaces.

Moreover, recalling (cf. Sect. 3.2.1) that every $f \in [\{0\}, GF(q)^*]\binom{n}{m}$, where
$GF(q)^*$ denotes the multiplicative group of $GF(q)$, represents an m-dimensional
linear subspace of $(GF(q))^n$ and that there is a bijective correspondence between
$[\{0\}, GF(q)^*]\binom{n}{1}$ and the one-dimensional linear subspaces of $(GF(q))^n$, the
Graham-Rothschild theorem (applied with $A = \{0\}$ and $G = GF(q)^*$) yields
a partition theorem for one-dimensional subspaces of linear spaces, i.e.,

Theorem 5.9. *Let $GF(q)$ be a finite field and m, r be positive integers. Then
there exists a positive integer $n = n(q, m, r)$ such that for every coloring Δ of
the one-dimensional linear subspaces of $(GF(q))^n$ with r colors there exists an*

m-dimensional linear subspace of $(GF(q))^n$ which is monochromatic with respect to Δ.   □

Observe that applying the Graham-Rothschild theorem with $A = GF(q)$ and G being the affine group acting on $GF(q)$, i.e., $G = \{\sigma \mid$ there exist $a, b \in GF(q), a \neq 0$, such that $\sigma(y) = ay + b$ for every $y \in GF(q)\ \}$, yields even a partition theorem for one-dimensional affine subspaces of $(GF(q))^n$.

Eventually, however, it was not an application of the Graham-Rothschild theorem which led to a proof of a general partition theorem for linear and affine spaces, i.e., to a resolution of Rota's conjecture, but an adaption of the methods used in the proof of this theorem.

Chapter 6
Canonical Partitions

Originally, Ramsey theory investigates the behavior of structures with respect to colorings of substructures into a fixed number of classes, typically into two classes. Probably the most well-known example is the pigeon hole principle, saying that for every 2-coloring of ω there exists an infinite subset $F \subseteq \omega$ which is monochromatic. Of course, if we allow colorings with an unbounded number of colors then it is clear that the conclusion of the pigeon hole principle does not have to hold. For example, we could take $\Delta(n) = n$ for every $n < \omega$. However, in this case we have an infinite set which meets each color in at most one element. Now it is an easy observation that one of these two possibilities must always occur. For every coloring $\Delta : \omega \to \omega$ there exists an infinite set $F \subseteq \omega$ such that either $\Delta \restriction F$ is monochromatic or $\Delta \restriction F$ is one-to-one, i.e., any two elements of F have different colors. This is the most elementary example of a *canonical partition theorem*, first introduced by Erdős and Rado (1950) studying unbounded colorings of finite sets. A coloring $\Delta : [\omega]^k \to \omega$ of the k-subsets of the nonnegative integers is canonical if there exists a $J \subseteq k$ such that $\Delta(X) = \Delta(Y)$ if and only if $X : J = Y : J$ for every pair $X, Y \in [\omega]^k$. The Erdős-Rado canonization Theorem 1.4 then asserts that for every coloring $\Delta : [\omega]^k \to \omega$ there exists $F \in [\omega]^\omega$ such that $\Delta \restriction [F]^k$ is canonical.

In this chapter unrestricted colorings of parameter words are investigated and their canonical patterns are determined. As applications we derive a canonizing version of van der Waerden's theorem from the corresponding result for zero-parameter words and the finite form of the Erdős-Rado canonization theorem from a canonizing version of the Graham-Rothschild theorem. In fact, throughout this chapter we will consider only parameter words over the trivial group.

A final remark concerns our notation. To indicate that we consider unbounded colorings we will always choose ω as their range, although it will quite often happen that only finitely many colors can actually be used.

H.J. Prömel, *Ramsey Theory for Discrete Structures*,
DOI 10.1007/978-3-319-01315-2_6,
© Springer International Publishing Switzerland 2013

Fig. 6.1 The canonical
pattern on 3^2

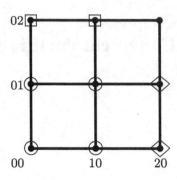

6.1 Canonizing Hales-Jewett's Theorem

In studying unbounded colorings of zero-parameter words we meet completely
different pattern of 'canonical colorings' than for finite sets. Consider, e.g., the
alphabet $3 = \{0, 1, 2\}$ and the equivalence relation $\approx$ on 3 having 0 and 1 in the
same class, 2 in another one. Define an (unbounded) coloring $\Delta_\approx : [3]\binom{n}{0} \to \omega$ by
$\Delta_\approx(g) = g/_\approx$, where $g/_\approx \in [\{0, 2\}]\binom{n}{0}$ is the $\approx$-quotient of g, i.e., $g/_\approx(i) = 0$
if $g(i) \in \{0, 1\}$, $g/_\approx(i) = 2$ otherwise. Observe that $\Delta_\approx$ obeys a kind of uniform
description. Any two m-parameter words inherit the same pattern from $\Delta_\approx$. In case
$m = 2$, i.e. of planes, this pattern can be visualized as in Fig. 6.1.

Of course, every equivalence relation on the alphabet $\{0, 1, 2\}$ leaves such a
hereditary pattern. More general, let A be any finite alphabet and let $\approx$ be an
equivalence relation on A. Then every coloring $\Delta_\approx : [A]\binom{n}{0} \to \omega$ satisfying

$$\Delta_\approx(g) = \Delta_\approx(h) \quad \text{if and only if} \quad g/_\approx = h/_\approx \tag{6.1}$$

is hereditary in the sense that for every m and every $f \in [A]\binom{n}{m}$ the restriction
$\Delta_\approx \rceil f \cdot A^m$ again satisfies (6.1). The following theorem shows that these are all
'canonical colorings'.

Theorem 6.1 (Canonical Hales-Jewett theorem). *Let A be a finite alphabet and
m be a positive integer. Then there exists a positive integer $n = CHJ(|A|, m)$ such
that for every unbounded coloring $\Delta : [A]\binom{n}{0} \to \omega$ there exists $f \in [A]\binom{n}{m}$ and
there exists an equivalence relation $\approx$ on A such that for all $g, h \in [A]\binom{m}{0}$ it follows
that*

$$\Delta(f \cdot g) = \Delta(f \cdot h) \quad \text{if and only if} \quad g/_\approx = h/_\approx,$$

$$\text{i.e., } g(i) \approx h(i) \text{ for every } i < m.$$

Observe that considering unbounded colorings we are only interested in the pattern of these colorings but not in the actual colors. This is taken into account by considering equivalence relations, thus abstracting from the actual colors.

A set $\mathcal{E}$ of equivalence relations on $[A]\binom{m}{k}$ is called a *canonical set of equivalence relations* if $\mathcal{E}$ is minimal (with respect to cardinality) such that there exists n so that for every unbounded coloring $\Delta : [A]\binom{n}{k} \to \omega$ there exists $f \in [A]\binom{n}{m}$ and an equivalence relation $\approx$ in $\mathcal{E}$ satisfying $\Delta(f \cdot g) = \Delta(f \cdot h)$ if and only if $h \approx g$, i.e., the equivalence relation induced by Δ coincides on f with $\approx$.

Theorem 6.1 together with the hereditary property of each of these equivalence relations imply that the set of all equivalence relations on $[A]\binom{m}{0}$ which are induced by equivalence relations on A, form a canonical set of equivalence relations on $[A]\binom{m}{0}$. In fact, this is the unique canonical set of equivalence relations on $[A]\binom{m}{0}$. Hence, it is justified to call a coloring $\Delta : [A]\binom{m}{0} \to \omega$ satisfying $\Delta(g) = \Delta(h)$ if and only if $g/\approx = h/\approx$ for some equivalence relation $\approx$ on A, a *canonical coloring of zero-parameter words*.

Proof of Theorem 6.1. Assume that $\Delta : [A]\binom{n}{0} \to \omega$ is given. Consider the colorings that a line $g \in [A]\binom{n}{1}$ induces: $\langle \Delta(g \cdot a) \mid a \in A \rangle$. In the following we are not interested in the actual coloring of the line, but only in its *pattern*, i.e., for which a's in A we get the same color and for which different ones. We can thus describe the pattern of a line by an equivalence relation on the alphabet A. Let r_a denote the number of equivalence relations on A. We just convinced ourselves that every coloring $\Delta : [A]\binom{n}{0} \to \omega$ gives rise to a coloring $\Delta^* : [A]\binom{n}{1} \to r_a$ which assigns to each line the equivalence relation on A that corresponds to the pattern induced by Δ on this line. Observe that the Graham-Rothschild theorem implies that for $n = GR(|A|, 1, M, r_a)$, where M is yet to be determined, there exists an $f \in [A]\binom{n}{M}$ that is monochromatic with respect to Δ^*.

We now repeat the above argument for m-spaces instead of lines. Every m-space $g \in [A]\binom{M}{m}$ induces a *pattern* with respect to the colors $\langle \Delta((f \cdot g) \cdot h) \mid h \in [A]\binom{m}{0} \rangle$ – and thus an equivalence relation on $[A]\binom{m}{0}$. Let $\hat{r}_a$ denote the number of equivalence relations on $[A]\binom{m}{0}$ and let $\Delta^{**} : [A]\binom{M}{m} \to \hat{r}_a$ denote the coloring that assigns to every m-space the equivalence relation on $[A]\binom{m}{0}$ that corresponds to the pattern induced by Δ on this m-space. Applying the Graham-Rothschild theorem again implies that for $M = GR(|A|, m, m + 1, \hat{r}_a)$ there exists a $f' \in [A]\binom{M}{m+1}$ that is monochromatic with respect to Δ^{**}.

Observe that f, f' induce a coloring $\hat{\Delta} : [A]\binom{m+1}{0} \to \omega$, defined by

$$\hat{\Delta}(\hat{f}) = \Delta((f \cdot f') \cdot \hat{f}) \quad \text{for every} \quad \hat{f} \in [A]\binom{m+1}{0}.$$

By construction we also have

(1) The pattern which $\hat{\Delta}$ leaves to lines are all the same, i.e.

$$\hat{\Delta}(\eta \cdot a) = \hat{\Delta}(\eta \cdot b) \quad \text{if and only if} \quad \hat{\Delta}(\eta' \cdot a) = \hat{\Delta}(\eta' \cdot b)$$

for all $\eta, \eta' \in [A]\binom{m+1}{1}$ and all $a, b \in A$,

and, additionally,

(2) The pattern which $\hat{\Delta}$ leaves to m-spaces are all the same, i.e.

$$\hat{\Delta}(\xi \cdot g) = \hat{\Delta}(\xi \cdot h) \quad \text{if and only if} \quad \hat{\Delta}(\xi' \cdot g) = \hat{\Delta}(\xi' \cdot h)$$

for all $\xi, \xi' \in [A]\binom{m+1}{m}$ and all $g, h \in [A]\binom{m}{0}$.

(We remark in passing that by repeating the above argument multiple times we could even ensure that the pattern which $\hat{\Delta}$ leaves to i-spaces are all the same – for all $1 \leq i \leq m$. However, in the following we do not need this generalization.)

In the following we use the notation $\overset{\Delta}{=}$ to abbreviate facts (1) and (2). More precisely, for $a, b \in A$ we write $a \overset{\Delta}{=} b$ if $\hat{\Delta}(\eta \cdot a) = \hat{\Delta}(\eta \cdot b)$ for some (and hence for all) $\eta \in [A]\binom{m+1}{1}$. Similarly, for $g, h \in [A]\binom{m}{0}$ we write $g \overset{\Delta}{=} h$ if $\hat{\Delta}(\xi \cdot g) = \hat{\Delta}(\xi \cdot h)$ for some (and hence again for all) $\xi \in [A]\binom{m+1}{m}$.

We define the relation $\approx$ on A as follows: $a \approx b$ if and only if $a \overset{\Delta}{=} b$. The idea now is to show that $g \overset{\Delta}{=} h$ if and only if $g/_{\approx} = h/_{\approx}$. Observe that in this case an m-parameter word $f \cdot f' \cdot \xi \in [A]\binom{n}{m}$, where $\xi \in [A]\binom{m+1}{m}$ is an arbitrarily chosen m-parameter word, together with $\approx$ satisfy the theorem.

First consider $g, h \in [A]\binom{m}{0}$ such that $g/_{\approx} = h/_{\approx}$. We prove by induction that

$$(g_0, g_1, \ldots, g_{m-1}) \overset{\Delta}{=} (h_0, h_1, \ldots, h_{i-1}, g_i, \ldots, g_{m-1})$$

for all $i \leq m$. This is trivially satisfied for $i = 0$. Assume it holds for some $i < m$, and consider the line $\eta = (h_0, h_1, \ldots, h_{i-1}, \lambda_0, g_{i+1} \ldots, g_{m-1}) \in [A]\binom{m}{1}$ and an arbitrary m-parameter word $\xi \in [A]\binom{m+1}{m}$. Observe that $g_i \approx h_i$ implies $\hat{\Delta}((\xi \cdot \eta) \cdot g_i) = \hat{\Delta}((\xi \cdot \eta) \cdot h_i)$, and thus $\eta \cdot g_i \overset{\Delta}{=} \eta \cdot h_i$. As

$$\eta \cdot g_i = (h_0, h_1, \ldots, h_{i-1}, g_i, \ldots, g_{m-1}) \qquad \text{and}$$
$$\eta \cdot h_i = (h_0, h_1, \ldots, h_{i-1}, h_i, g_{i+1} \ldots, g_{m-1}),$$

we deduce that the induction hypothesis also holds for $i + 1$. Note that for $i = m$ we get $g \overset{\Delta}{=} h$, as desired.

Let us now assume that $g, h \in [A]\binom{m}{0}$ are such that $g/_{\approx} \neq h/_{\approx}$. Choose $i \in m$ with $g_i \not\approx h_i$ and consider $\eta = (g_0, \ldots, g_{i-1}, g_i, \lambda_0, g_{i+1}, \ldots, g_{m-1}) \in [A]\binom{m+1}{1}$. Then $g_i \not\approx h_i$ implies that

$$\hat{\Delta}(g_0,\dots,\dots,g_{i-1},g_i,g_i,g_{i+1},\dots,g_{m-1}) = \hat{\Delta}(\eta \cdot g_i)$$

$$\neq \hat{\Delta}(\eta \cdot h_i) = \hat{\Delta}(g_0,\dots,\dots,g_{i-1},g_i,h_i,g_{i+1},\dots,g_{m-1}). \quad (6.2)$$

In order derive a contradiction assume that $g \stackrel{\Delta}{\equiv} h$ and consider m-parameter words

$$\xi = (\lambda_0,\dots,\lambda_{i-1},\lambda_i,\lambda_i,\lambda_{i+1},\dots,\lambda_{m-1}),$$

$$\xi' = (\lambda_0,\dots,\lambda_{i-1},\lambda_i,h_i,\lambda_{i+1},\dots,\lambda_{m-1}) \in [A]\binom{m+1}{m}.$$

Then $g \stackrel{\Delta}{\equiv} h$ implies $\hat{\Delta}(\xi \cdot g) = \hat{\Delta}(\xi \cdot h)$ and $\hat{\Delta}(\xi' \cdot g) = \hat{\Delta}(\xi' \cdot h)$. Closer inspection of the words ξ and ξ' yields that $\xi \cdot h = \xi' \cdot h$, thus

$$\hat{\Delta}(g_0,\dots,\dots,g_{i-1},g_i,g_i,g_{i+1},\dots,g_{m-1}) = \hat{\Delta}(\xi \cdot g)$$

$$= \hat{\Delta}(\xi' \cdot g) = \hat{\Delta}(g_0,\dots,\dots,g_{i-1},g_i,h_i,g_{i+1},\dots,g_{m-1}).$$

which contradicts (6.2). Hence $g \stackrel{\Delta}{\not\equiv} h$, as desired. This completes the proof of Theorem 6.1.

Schmerl (1993) applies this result to show that for every countable non-standard model $\mathcal{M}$ of Peano arithmetic and every positive integer $k \geq 2$ there exists a cofinal extension $\mathcal{N}$ of $\mathcal{M}$ such that the lattice $\mathcal{L}(\mathcal{N}/\mathcal{M})$ of intermediate models is isomorphic to $\Pi(k)$, the lattice of equivalence relations of a k-element set (cf. also Schmerl 1985).

The special case $|A| = 2$ of Theorem 6.1 admits the following formulation.

Corollary 6.2. *Let m be a positive integer. Then there exists a positive integer $n = CHJ(2, m)$ such that for every coloring $\Delta : \mathcal{B}(n) \to \omega$ of the points of the n-dimensional Boolean lattice $\mathcal{B}(n)$ there exists a $\mathcal{B}(m)$-sublattice $\mathcal{L} \subseteq \mathcal{B}(n)$ such that either $\Delta \rceil \mathcal{L}$ is constant or $\Delta \rceil \mathcal{L}$ is one-to-one.* $\square$

Here we have the same kind of result as for the unbounded pigeon hole principle: the substructure we are looking for must either be colored monochromatically or one-to-one. Nešetřil and Rödl (1978b, 1979) call this phenomenon *selectivity*. We will meet this phenomenon several times in the sequel, e.g., in the next section in connection with van der Waerden's theorem.

Recall that every finite poset can be embedded (order-preservingly) into some Boolean lattice $\mathcal{B}(n)$, cf. Sect. 4.2.5. Hence, we get immediately

Corollary 6.3. *Let Q be a finite poset. Then there exists a finite poset P such that for every coloring $\Delta : P \to \omega$ of the points of P there exists a Q-subposet $Q' \subseteq P$ such that either $\Delta \rceil Q'$ is monochromatic or $\Delta \rceil Q'$ is one-to-one.* $\square$

6.2 Canonizing van der Waerden's Theorem

As indicated in Sect. 4.2, van der Waerden's theorem on arithmetic progressions is one of the most prominent applications of Hales-Jewett's theorem. The aim of this section is to show how a canonical version of van der Waerden's theorem can be obtained using the canonical Hales-Jewett theorem.

Theorem 6.4 (Canonical van der Waerden theorem). *Let t be a positive integer. Then there exists a positive integer $n = EG(t)$ such that for every coloring Δ : $n \to \omega$ there exists a t-term arithmetic progression $X \subseteq n$ such that either $\Delta \rceil X$ is constant or $\Delta \rceil X$ is one-to-one.*

At the first glance it may look somewhat astonishing that the canonical Hales-Jewett theorem which allows every pattern on the lines can be used in order to obtain a selectivity result for arithmetic progressions. The original proof of Erdős and Graham (1980) used Szemerédi's density result for arithmetic progressions. Later, an 'elementary' proof was obtained by Prömel and Rödl (1986). The proof given here is based on ideas from (Prömel and Rothschild 1987) which can also be used to prove a slightly stronger result, viz. a restricted version of the canonical van der Waerden theorem.

Proof of Theorem 6.4. Let $\ell = (t - 1)^2 + 1$. It is easy to see that the first ℓ nonnegative integers have the following property:

(1) Let $\mu < \nu < t$ be arbitrary and let $\approx$ be an equivalence relation on ℓ such that every arithmetic progression of length t in ℓ has its μth and its νth term in the same equivalence class. Then there is a t-term arithmetic progression in ℓ which is completely contained in one equivalence class, e.g., the progression $\mu + (\nu - \mu) \cdot j, j < t$.

Let $(X_i)_{i<z}$ be an enumeration of all arithmetic progressions of length t in ℓ and assume $X_i = \{x_{i,0}, \ldots, x_{i,t-1}\}$ for every $i < z$ is in ascending order.

Choose $n = CHJ(\ell, z)$ according to the canonical Hales-Jewett theorem and let $\Delta : (\ell - 1)n + 1 \to \omega$ be a coloring. Consider the coloring $\Delta^* : [\ell]\binom{n}{0} \to \omega$ which is defined by

$$\Delta^*(g_0, \ldots, g_{n-1}) = \Delta(\textstyle\sum_{i<n} g_i).$$

By choice of n there exists $f \in [\ell]\binom{n}{z}$ and an equivalence relation $\approx$ on ℓ such that $\Delta^* \rceil f$ is canonical, meaning that for all $g, h \in [\ell]\binom{z}{0}$ we have:

$$\Delta^*(f \cdot g) = \Delta^*(f \cdot h) \quad \text{if and only if} \quad g_i \approx h_i \text{ for every } i < z.$$

Let $F = \sum\{f_i \mid f_i \in \ell\}$ and put $\zeta^j = (x_{0,j}, x_{1,j}, \ldots, x_{z-1,j})$ for $j < t$ and consider $\{f \cdot \zeta^j \mid j < t\}$. Observe that $\{F + \sum_{i<z} x_{ij} \mid j < t\}$ forms a t-term arithmetic progression.

First assume that $\Delta^*]\{f \cdot \zeta^j \mid j < t\}$ is one-to-one. Then, clearly, $\Delta]\{F + \sum_{i<z} x_{ij} \mid j < t\}$ is also one-to-one and we are done.

So assume that there exists $\mu, \nu < t$ such that

$$\Delta^*(f \cdot \zeta^\mu) = \Delta^*(f \cdot \zeta^\nu).$$

But then $x_{j,\mu} \approx x_{j,\nu}$ for every $j < z$. So by (1) there exists an arithmetic progression X_i such that $x_{i,0} \approx x_{i,1} \approx \ldots \approx x_{i,t-1}$. Let

$$\xi^j = (\underbrace{0, \ldots, 0}_{z-1}, x_{ij}),$$

for every $j < t$. Then $\Delta^*]\{f \cdot \xi^j \mid j < t\}$ is constant and hence, by definition, also $\Delta]\{F + x_{ij} \mid j < t\}$. Observing that $\{F + x_{ij} \mid j < t\}$ forms a t-term arithmetic progression completes the proof of Theorem 6.4.   □

Concerning more than one dimension a canonical version of Gallai-Witt's theorem was proved by Deuber et al. (1983) for finite subsets of the integer lattice grid and by Spencer (1983) for arbitrary finite subsets of the Euclidean space, both based on Fürstenberg-Katznelson's density version of the Gallai-Witt result. Simplified proofs are given in Prömel and Rödl (1986) and Prömel and Rothschild (1987). Although the method used to prove the canonical van der Waerden theorem can easily be adopted to derive a canonical version of Gallai-Witt's theorem there exist additional canonical patterns in this higher dimensional case. We omit the result.

6.3 Canonizing Graham-Rothschild's Theorem

Next we consider an extension of the canonizing version of Hales-Jewett's theorem to higher dimensions. Here, the canonical colorings occurring in the Erdős-Rado canonization theorem and those from the canonical Hales-Jewett theorem come together, finding a kind of common generalization.

Consider the surjective mapping $\phi : [A]\binom{m}{k} \to [m]^k$ given by $\phi(f) = \{\min f^{-1}(\lambda_i) \mid i < k\}$, cf. Sect. 3.1.2. This mapping shows that every canonical coloring $\Delta_J : [m]^k \to \omega$, where $J \subseteq k$ and $\Delta_J(X) = X : J$, gives rise to a canonical coloring

$$\Delta : [A]\binom{m}{k} \to \omega \quad \text{via} \quad \Delta(f) = (\phi(f)) : J.$$

On the other hand, every equivalence relation $\approx$ on $A \cup \{\lambda_0, \ldots, \lambda_{k-1}\}$ allows to color according to the $\approx$-quotient of the k-parameter words in $[A]\binom{m}{k}$.

It turns out that all colorings which are relevant for canonizing the Graham-Rothschild theorem can be produced by combining these two types of colorings appropriately.

Let $J \subseteq k$ be any subset of k and put $J^+ = J \cup \{k\}$. For $i \in k$ let $pre(i) := \max\{j \in J^+ \mid j < i\}$ (and $pre(i) = -1$ if there doesn't exist such element in J), and $suc(i) := \min\{j \in J^+ \mid j > i\}$. Consider a family of equivalence relations $\{\approx_i\}_{i \in J^+}$, where $\approx_i$ is defined on $A \cup \{\lambda_0, \ldots, \lambda_{i-1}\}$. We associate to the pair $\Pi = (J, (\approx_i)_{i \in J^+})$ an equivalence relation $\approx_\Pi$ on $[A]\binom{n}{k}$ by putting

$g \approx_\Pi h$ if and only if   for every $i \in J^+$

$\quad$ (1) $\quad \min g^{-1}(\lambda_i) = \min h^{-1}(\lambda_i)$,

$\quad$ (2) $\quad g(v) \approx_i h(v) \; \forall \; \min g^{-1}(\lambda_{pre(i)}) < v < \min g^{-1}(\lambda_i)$,

where we tacitly agree that $\min g^{-1}(\lambda_{-1}) = -1$ and $\min g^{-1}(\lambda_k) = m$.

Note that the definition of $\approx_\Pi$ does not depend on the dimension of the parameter words on which it is imposed. The pair $\Pi = (J, (\approx_i)_{i \in J^+})$ is called an (A, k)-*canonical pair*, if and only if

(3) For every $j \in J$ we have $\alpha \approx_j \beta$ implies $\alpha \approx_{suc(j)} \beta$ for all $\alpha, \beta \in A \cup \{\lambda_0, \ldots, \lambda_{j-1}\}$, i.e., the family of equivalence relations is getting coarser, and

(4) For every $j \in \{0, \ldots, k-1\} \setminus J$ there exists $\alpha \in A \cup \{\lambda_0 \ldots, \lambda_{j-1}\}$ such that $\alpha \approx_{suc(j)} \lambda_j$.

Observe that condition (3) assures that the associated equivalence relation $\approx_\Pi$ is hereditary, meaning that for every $f \in [A]\binom{n}{m}$ the restriction of $\approx_\Pi$ to f yields the same equivalence relation, i.e., $f \cdot g \approx_\Pi f \cdot h$ if and only if $g \approx_\Pi h$. We prove now that any two equivalence relations which are associated to distinct canonical pairs are essentially different and then we show that the set of equivalence relations which come from (A, k)-canonical pairs indeed forms a canonical set of equivalence relations on $[A]\binom{n}{k}$.

Proposition 6.5. *Let* $\Pi_0 = (J_0, (\approx_i^0)_{i \in J_0^+})$ *and* $\Pi_1 = (J_1, (\approx_i^1)_{i \in J_1^+})$ *be distinct* (A, k)-*canonical pairs. Then for every* $f \in [A]\binom{n}{m}$ *the restrictions of* $\approx_{\Pi_0}$ *and* $\approx_{\Pi_1}$ *to* f *are distinct.*

Proof. Fix some $f \in [A]\binom{n}{m}$. First assume that $J_0 \neq J_1$. Without loss of generality we can assume that there exists $j \in J_0$ such that $j \notin J_1$. By (4) we know that there exists $\alpha \in A \cup \{\lambda_0, \ldots, \lambda_{j-1}\}$ so that $\alpha \approx_i^1 \lambda_j$ (where $i > j$ is minimal so that $i \in J_1^+$). Consider

$$g = (\lambda_0, \ldots, \lambda_{j-1}, \lambda_j, \lambda_j, \lambda_{j+1}, \ldots, \lambda_{k-1}, \lambda_0, \ldots, \lambda_0) \in [A]\binom{m}{k}$$

and

$$h = (\lambda_0, \ldots, \lambda_{j-1}, \alpha, \lambda_j, \lambda_{j+1}, \ldots, \lambda_{k-1}, \lambda_0, \ldots, \lambda_0) \in [A]\binom{m}{k}.$$

Then,

$$f \cdot g \not\approx_{\Pi_0} f \cdot h, \quad \text{as } \min(f \cdot g)^{-1}(\lambda_j) \neq \min(f \cdot h)^{-1}(\lambda_j), \quad \text{but}$$

$$f \cdot g \approx_{\Pi_1} f \cdot h, \quad \text{as } \alpha \approx_i^1 \lambda_j \text{ implies by (3) that } \alpha \approx_\ell^1 \lambda_j \text{ for every } i \leq \ell \leq k.$$

Now assume that $J_0 = J_1$, but there exist $i \in J_0^+$ and $\alpha, \beta \in A \cup \{\lambda_0, \ldots, \lambda_{i-1}\}$ so that $\alpha \not\approx_i^0 \beta$, but $\alpha \approx_i^1 \beta$. Put

$$g = (\lambda_0, \ldots, \lambda_{i-1}, \alpha, \lambda_i, \lambda_{i+1}, \ldots, \lambda_{k-1}, \lambda_0, \ldots, \lambda_0) \in [A]\binom{m}{k}$$

and

$$h = (\lambda_0, \ldots, \lambda_{i-1}, \beta, \lambda_i, \lambda_{i+1}, \ldots, \lambda_{k-1}, \lambda_0, \ldots, \lambda_0) \in [A]\binom{m}{k}.$$

Then, obviously, $f \cdot g \not\approx_{\Pi_0} f \cdot h$, but $f \cdot g \approx_{\Pi_1} f \cdot h$, as above, completing the proof of Proposition 6.5. $\qquad \square$

Theorem 6.6 (Canonical Graham-Rothschild theorem). *Let A be a finite alphabet and k, m be positive integers. Then there exists $n = PV(|A|, k, m)$ such that for every coloring $\Delta : [A]\binom{n}{k} \to \omega$ there exists $f \in [A]\binom{n}{m}$ and there exists an (A, k)-canonical pair $\Pi = (J, (\approx_i)_{i \in J^+})$ such that for all $g, h \in [A]\binom{m}{k}$ we have*

$$\Delta(f \cdot g) = \Delta(f \cdot h) \quad \text{if and only if} \quad g \approx_\Pi h.$$

Proof. Proceeding as in the proof of the canonical Hales-Jewett theorem we observe that by using the (classical) Graham-Rothschild theorem twice we may assume that there exists $\hat{f} \in [A]\binom{n}{m+1}$ such that $\hat{\Delta} : [A]\binom{m+1}{k} \to \omega$,

$$\hat{\Delta}(g) := \Delta(\hat{f} \cdot g) \quad \text{for} \quad g \in [A]\binom{m+1}{k},$$

satisfies:

(1a) The pattern which $\hat{\Delta}$ leaves to the $(k+1)$-parameter subwords are all the same, i.e.,

$$\hat{\Delta}(\eta \cdot a) = \hat{\Delta}(\eta \cdot b) \quad \text{if and only if} \quad \hat{\Delta}(\eta' \cdot a) = \hat{\Delta}(\eta' \cdot b)$$

for all $\eta, \eta' \in [A]\binom{m+1}{k+1}$ and all $a, b \in [A]\binom{k+1}{k}$, and additionally,

(1b) The pattern which $\hat{\Delta}$ leaves to the m-parameter subwords are all the same, i.e.,

$$\hat{\Delta}(\xi \cdot a) = \hat{\Delta}(\xi \cdot b) \quad \text{if and only if} \quad \hat{\Delta}(\xi' \cdot a) = \hat{\Delta}(\xi' \cdot b)$$

for all $\xi, \xi' \in [A]\binom{m+1}{m}$ and all $a, b \in [A]\binom{m}{k}$

We define the relation $\stackrel{\triangle}{=}$ similarly as in the proof of the canonical Hales-Jewett theorem: for $t \in \{k+1, m\}$ and $a, b \in [A]\binom{t}{k}$, we write $a \stackrel{\triangle}{=} b$ if $\hat{A}(f' \cdot a) = \hat{A}(f' \cdot b)$ for some (and hence for all) $f' \in [A]\binom{m+1}{t}$. We also extend this notation to other values $t \in \{k+1, \ldots, m+1\}$ as follows: for $a, b \in [A]\binom{t}{k}$, we write $a \stackrel{\triangle}{=} b$ if $\hat{A}(f' \cdot a) = \hat{A}(f' \cdot b)$ for all $f' \in [A]\binom{m+1}{t}$. We will repeatedly make use of the following simple fact that shows that the relation $\stackrel{\triangle}{=}$ can be extended upwards:

(1c) If $a \stackrel{\triangle}{=} b$ for some $a, b \in [A]\binom{t}{k}$, then $f'' \cdot a \stackrel{\triangle}{=} f'' \cdot b$ for every $f'' \in [A]\binom{t'}{t}$, $t' \in \{t, \ldots, m+1\}$.

To see this fix some $f'' \in [A]\binom{t'}{t}$ and consider an arbitrary $f''' \in [A]\binom{m+1}{t'}$; then $f''' \cdot f'' \in [A]\binom{m+1}{t}$ and $a \stackrel{\triangle}{=} b$ thus implies that $\hat{A}(f''' \cdot f'' \cdot a) = \hat{A}(f''' \cdot f'' \cdot b)$.

It remains to find an (A, k)-canonical pair Π such that

$$g \stackrel{\triangle}{=} h \quad \text{if and only if} \quad g \approx_\Pi h,$$

for every pair $g, h \in [A]\binom{m}{k}$. Note that then Π together with an m-parameter word $\hat{f} \cdot f$, where $f \in [A]\binom{m+1}{m}$ is chosen arbitrarily, satisfies the theorem.

First we define equivalence relations $\approx_i^*$ for all $i \leq k$. These equivalence relations will later be used to obtain a set $J \subseteq k$ and a family of equivalence relations $\approx_i$, $i \in J^+$, which form an (A, k)-canonical pair. Let $\approx_i^*$ be defined on $A \cup \{\lambda_0, \ldots, \lambda_i\}$, by $\alpha \approx_i^* \beta$ if and only if $\lambda^i(\alpha) \stackrel{\triangle}{=} \lambda^i(\beta)$, where

$$\lambda^i(x) = (\lambda_0, \ldots, \lambda_{i-1}, x, \lambda_i, \ldots, \lambda_{k-1}).$$

In order to later define the desired set J, we first exhibit three properties of the relations $\approx_i^*$:

(2a) $\alpha \approx_i^* \beta$ implies $\alpha \approx_{i+1}^* \beta$, thus $\approx_{i+1}^*$ is coarser than $\approx_i^*$, for every $i < k$.

(2b) Let $\alpha \approx_i^* \lambda_i$ for some $\alpha \in A \cup \{\lambda_0, ..., \lambda_{i-1}\}$. Then $\approx_{i+1}^*]$ $A \cup \{\lambda_0, \ldots, \lambda_i\} = \approx_i^*$.

Every parameter word $g \in [A]\binom{m}{k}$ is naturally divided into $k+1$ (possibly empty) pieces between the minimal occurrences of its k parameters. We denote by $p(g, i) \subseteq m$ the positions of the ith of these $k+1$ pieces. More formally,

$$p(g, i) = \{j < m \mid \min g^{-1}(\lambda_{i-1}) < j < \min g^{-1}(\lambda_i)\}$$

for $i < k$, where we assume that $\min g^{-1}(\lambda_{-1}) = -1$ and $\min g^{-1}(\lambda_k) = m$.

(2c) Let $g \in [A]\binom{m}{k}$ and $\ell \in m$ such that $\ell \in p(g, i)$ for some $i \in k+1$. Then for any $\alpha \in A \cup \{\lambda_0, \ldots, \lambda_i\}$ such that $g_\ell \approx_i^* \alpha$ and

$$g' = (g_0, \ldots, g_{\ell-1}, \alpha, g_{\ell+1}, \ldots, g_{m-1}) \in [A]\binom{m}{k}$$

we have $g \stackrel{\triangle}{=} g'$.

Proof of (2a): Assume that $\alpha \approx_i^* \beta$. Applying (1c) with

$$\eta = (\lambda_0, \ldots, \lambda_i, \lambda_{i+1}, \lambda_i, \lambda_{i+2}, \ldots, \lambda_k),$$
$$\eta' = (\lambda_0, \ldots, \lambda_i, \lambda_{i+1}, \alpha, \lambda_{i+2}, \ldots, \lambda_k) \in [A]\binom{k+2}{k+1}$$

on $\lambda^i(\alpha) \triangleq \lambda^i(\beta)$, we get

$$\eta \cdot \lambda^i(\alpha) = (\lambda_0, \ldots, \lambda_{i-1}, \alpha, \lambda_i, \alpha, \lambda_{i+1}, \ldots, \lambda_{k-1})$$
$$\triangleq (\lambda_0, \ldots, \lambda_{i-1}, \beta, \lambda_i, \beta, \lambda_{i+1}, \ldots, \lambda_{k-1}) = \eta \cdot \lambda^i(\beta),$$

and

$$\eta' \cdot \lambda^i(\alpha) = (\lambda_0, \ldots, \lambda_{i-1}, \alpha, \lambda_i, \alpha, \lambda_{i+1}, \ldots, \lambda_{k-1})$$
$$\triangleq (\lambda_0, \ldots, \lambda_{i-1}, \beta, \lambda_i, \alpha, \lambda_{i+1}, \ldots, \lambda_{k-1}) = \eta' \cdot \lambda^i(\beta).$$

Thus, by transitivity,

$$(\lambda_0, \ldots, \lambda_{i-1}, \beta, \lambda_i, \beta, \lambda_{i+1}, \ldots, \lambda_{k-1}) \triangleq (\lambda_0, \ldots, \lambda_{i-1}, \beta, \lambda_i, \alpha, \lambda_{i+1}, \ldots, \lambda_{k-1}).$$

Now consider $\eta'' = (\lambda_0, \ldots, \lambda_{i-1}, \beta, \lambda_i, \ldots, \lambda_k) \in [A]\binom{k+2}{k+1}$ and observe that the equality above implies

$$\eta'' \cdot \lambda^{i+1}(\alpha) = (\lambda_0, \ldots, \lambda_{i-1}, \beta, \lambda_i, \alpha, \lambda_{i+1}, \ldots, \lambda_{k-1})$$
$$\triangleq (\lambda_0, \ldots, \lambda_{i-1}, \beta, \lambda_i, \beta, \lambda_{i+1}, \ldots, \lambda_{k-1})$$
$$= \eta'' \cdot \lambda^{i+1}(\beta).$$

Therefore, for any $f \in [A]\binom{m+1}{k+2}$ we have $\hat{\Delta}((f \cdot \eta'') \cdot \lambda^{i+1}(\alpha)) = \hat{\Delta}((f \cdot \eta'') \cdot \lambda^{i+1}(\beta))$, hence from (1a) we deduce $\lambda^{i+1}(\alpha) \triangleq \lambda^{i+1}(\beta)$ which by definition implies $\alpha \approx_{i+1}^* \beta$, proving (2a).

Proof of (2b): Let us assume $\alpha \approx_i^* \lambda_i$ for some $\alpha \in A \cup \{\lambda_0, \ldots, \lambda_{i-1}\}$. From (2a) we already know that $\approx_{i+1}^*$ is coarser than $\approx_i^*$. We need to show that $\approx_{i+1}^*$ restricted to $A \cup \{\lambda_0, \ldots, \lambda_i\}$ is not strictly coarser than $\approx_i^*$. In other words, we need to show that for $\beta, \gamma \in A \cup \{\lambda_0, \ldots, \lambda_i\}$ with $\beta \approx_{i+1}^* \gamma$ we also have $\beta \approx_i^* \gamma$. Observe that the assumption $\alpha \approx_i^* \lambda_i$ implies that $\alpha \approx_{i+1}^* \lambda_i$ (as $\approx_{i+1}^*$ is coarser). That is, without loss of generality we may assume that neither β nor γ is equal to λ_i.

We proceed similarly as in the proof of (2a). Applying (1c) with

$$\eta = (\lambda_0, \ldots, \lambda_{i-1}, \lambda_i, \beta, \lambda_{i+1}, \ldots, \lambda_k),$$
$$\eta' = (\lambda_0, \ldots, \lambda_{i-1}, \lambda_i, \gamma, \lambda_{i+1}, \ldots, \lambda_k) \in [A]\binom{k+2}{k+1}$$

on $\lambda^i(\alpha) \triangleq \lambda^i(\lambda_i)$, we get

$$(\lambda_0, \ldots, \lambda_{i-1}, \alpha, \beta, \lambda_i, \ldots, \lambda_{k-1}) \triangleq (\lambda_0, \ldots, \lambda_{i-1}, \lambda_i, \beta, \lambda_i, \ldots, \lambda_{k-1}),$$
$$(\lambda_0, \ldots, \lambda_{i-1}, \alpha, \gamma, \lambda_i, \ldots, \lambda_{k-1}) \triangleq (\lambda_0, \ldots, \lambda_{i-1}, \lambda_i, \gamma, \lambda_i, \ldots, \lambda_{k-1}).$$

Similarly, applying (1c) with

$$\eta'' = (\lambda_0, \ldots, \lambda_{i-1}, \lambda_i, \lambda_{i+1}, \lambda_i, \lambda_{i+2}, \ldots, \lambda_k) \in [A]\binom{k+2}{k+1}$$

on $\lambda^{i+1}(\beta) \triangleq \lambda^{i+1}(\gamma)$, which follows from $\beta \approx^*_{i+1} \gamma$, implies

$$(\lambda_0, \ldots, \lambda_i, \beta, \lambda_i, \lambda_{i+1}, \ldots, \lambda_{k-1}) \triangleq (\lambda_0, \ldots, \lambda_i, \gamma, \lambda_i, \lambda_{i+1}, \ldots, \lambda_{k-1}).$$

Therefore, by transitivity we have

$$(\lambda_0, \ldots, \lambda_{i-1}, \alpha, \beta, \lambda_i, \ldots, \lambda_{k-1}) \triangleq \Delta(\lambda_0, \ldots, \lambda_{i-1}, \alpha, \gamma, \lambda_i, \ldots, \lambda_{k-1}).$$

Thus, applying $\eta''' = (\lambda_0, \ldots, \lambda_{i-1}, \alpha, \lambda_i, \ldots, \lambda_k) \in [A]\binom{k+2}{k+1}$ on the previous equality, we have

$$\begin{aligned} \eta''' \cdot \lambda^i(\beta) &= (\lambda_0, \ldots, \lambda_{i-1}, \alpha, \beta, \lambda_i, \ldots, \lambda_{k-1}) \\ &\triangleq (\lambda_0, \ldots, \lambda_{i-1}, \alpha, \gamma, \lambda_i, \ldots, \lambda_{k-1}) \\ &= \eta''' \cdot \lambda^i(\gamma). \end{aligned}$$

By the same argument as in the proof of (2a) we deduce that $\beta \approx^*_i \gamma$, thus proving (2b).

Proof of (2c): Let $g \in [A]\binom{m}{k}$ and $\ell \in p(g, i)$ for some $i \in k+1$, and consider any $\alpha \in A \cup \{\lambda_0, \ldots, \lambda_i\}$ such that $\alpha \approx^*_i g_\ell$. Then by applying (1c) with

$$\eta = (g_0, \ldots, g_{\ell-1}, \lambda_i, g^*_{\ell+1}, \ldots, g^*_{m-1}) \in [A]\binom{m}{k+1},$$

where

$$g^*_v = \begin{cases} g_v & \text{if } g_v \in A \cup \{\lambda_0, \ldots, \lambda_{i-1}\} \\ \lambda_{\mu+1} & \text{if } g_v = \lambda_\mu \text{ for } \mu \geq i, \end{cases}$$

on $\lambda^i(\alpha) \triangleq \lambda^i(g_\ell)$, we get

$$g = \eta \cdot \lambda^i(g_\ell) \triangleq \eta \cdot \lambda^i(\alpha) = g',$$

which completes the proof of (2c).

Completing the proof: With properties (2a), (2b) and (2c) at hand we complete the proof of the theorem as follows. Let

$$J := \{i < k \mid \alpha \not\approx_i^* \lambda_i \text{ for every } \alpha \in A \cup \{\lambda_0, \ldots, \lambda_{i-1}\}\} \quad \text{and}$$

$$\approx_i := \approx_i^* {\restriction} A \cup \{\lambda_0, \ldots, \lambda_{i-1}\} \text{ for every } i \in J^+.$$

We now show that these $J \subseteq k$ and $\approx_i$, $i \in J^+$, are such that $\Pi = (J, (\approx_i)_{i \in J^+})$ is as required in the theorem.

By (2a) and the definition of J it is obvious that $\Pi = (J, (\approx_i)_{i \in J^+})$ is an (A, k)-canonical pair. In the remainder of the proof we verify that $g \triangleq h$ if and only if $g \approx_\Pi h$, for all $g, h \in [A]\binom{m}{k}$. In doing so we will repeatedly use the following observation which immediately follows from the definition of J and (2b):

(*) If $i < k$, $j \in J$ are such that $pre(j) < i \le j$, then $\pi_i^* = \pi_j {\restriction} A \cup \{\lambda_0, \ldots, \lambda_i\}$

First assume that $g \approx_\Pi h$. We show, by induction, that there exist k-parameter words $g^0, \ldots, g^k, h^0, \ldots, h^k \in [A]\binom{m}{k}$, such that for each $t \in k + 1$ the following holds:

(3a) $\min(g^t)^{-1}(\lambda_i) = \min(h^t)^{-1}(\lambda_i)$ for $i \in t$, i.e., the first occurrences of each of the first t parameters are identical in g^t and h^t.
(3b) $g^t \approx_\Pi h^t$, and
(3c) $g^{t-1} \triangleq g^t$ and $h^{t-1} \triangleq h^t$,

where $g^{-1} = g$ and $h^{-1} = h$. For $t = 0$, all three properties are trivially satisfied for $g^0 = g$ and $h^0 = h$. Assume now that the claim holds for some $t \in k$. If $\min(g^t)^{-1}(\lambda_t) = \min(h^t)^{-1}(\lambda_t)$, then $g^{t+1} = g^t$ and $h^{t+1} = h^t$ satisfies the claim for $t + 1$. Otherwise, without loss of generality we assume that $\min(g^t)^{-1}(\lambda_t) > \min(h^t)^{-1}(\lambda_t)$. Note that this implies $t \notin J$ (as $g^t \approx_\Pi h^t$) and $\ell = \min(h^t)^{-1}(\lambda_t) \in p(g^t, t)$. From (*) and $g^t \approx_\Pi h^t$ it thus follows that $g_\ell^t \approx_t^* h_\ell^t$. Thus by applying (2c) with $\alpha = h_\ell^t = \lambda_t$ on g^t we get $g^{t+1} \in [A]\binom{m}{k}$,

$$g^{t+1} = (g_0^t, \ldots, g_{\ell-1}^t, \lambda_t, g_{\ell+1}^t, \ldots, g_{m-1}^t),$$

such that $g^t \triangleq g^{t+1}$. It is easy now to see that g^{t+1}, together with $h^{t+1} = h^t$, satisfies all three properties of the claim. For $t = k$, (3a) implies that g^k and h^k agree on all first occurrences of parameters. Thus for each $\ell \in m$ such that $g_\ell^k \neq h_\ell^k$ we have $\ell \in p(g^k, i)$, for some $i \in k + 1$. Since $g^k \approx_\Pi h^k$, we can apply (2c) together with (*) for $\alpha = h_\ell^k$ on g^k, hence completely matching g^k and h^k. Therefore $g^k \triangleq h^k$, and from (3c) we conclude $g \triangleq h$.

Let us now assume that $g \not\approx_\Pi h$. First we show that we may assume without loss of generality that g and h are such that there exists a position ℓ and an index $i < k$ such that the following three properties are satisfied:

(4a) $g_\ell \not\approx_i^* h_\ell$,
(4b) For all $i' < i$ we have $\min g^{-1}(\lambda_{i'}) = \min h^{-1}(\lambda_{i'}) < \ell$,

(4c) $\ell \leq \min g^{-1}(\lambda_i) \leq \min h^{-1}(\lambda_i)$.

If the first occurrences of the parameters λ_j for $j < k$ are all identical, then $g \not\approx_\Pi h$ together with $(*)$ easily implies that there exist indices i and ℓ that satisfy (4a)–(4c). Otherwise choose $i < k$ minimal such that $\min g^{-1}(\lambda_i) \neq \min h^{-1}(\lambda_i)$. We may assume without loss of generality (rename g and h if necessary) that $\ell := \min g^{-1}(\lambda_i) < \min h^{-1}(\lambda_i)$. If $g_\ell \not\approx_i^* h_\ell$ then we have found ℓ and i that satisfy (4a)–(4c). So assume that $g_\ell \approx_i^* h_\ell$. Apply (2c) to deduce that $h' = (h_1, h_{\ell-1}, g_\ell, h_{\ell+1}, \ldots, h_{m-1})$ satisfies $h \triangleq h'$. Note that $(*)$ implies that we also have that $h' \approx_\Pi h$. We may thus assume without loss of generality that $h = h'$. Repeating this process we see that we either find the desired ℓ and i or we end up with g and h such that for all $i < k$ we have $\min g^{-1}(\lambda_i) = \min h^{-1}(\lambda_i)$, which is the case that we already handled.

So assume now that ℓ and i are such that (4a)–(4c) hold. Consider the $(k+1)$-parameter word $\eta \in [A]\binom{m+1}{k+1}$,

$$\eta = (g_0, \ldots, g_{\ell-1}, \lambda_i, g_\ell^*, \ldots, g_{m-1}^*),$$

where

$$g_\nu^* = \begin{cases} g_\nu & \text{if } g_\nu \in A \cup \{\lambda_0, \ldots, \lambda_{i-1}\} \\ \lambda_{j+1} & \text{if } g_\nu = \lambda_j \text{ for } j \geq i. \end{cases}$$

Note that $g_\ell \not\approx_i^* h_\ell$ implies, by definition, $\lambda^i(g_\ell) \not\triangleq \lambda^i(h_\ell)$. Then from (1a) we also have $\eta \cdot \lambda^i(g_\ell) \not\triangleq \eta \cdot \lambda^i(h_\ell)$, thus

$$\eta \cdot \lambda^i(g_\ell) = (g_0, \ldots, g_{\ell-1}, g_\ell, g_\ell, \ldots, g_{m-1}) \qquad (6.3)$$

$$\not\triangleq (g_0, \ldots, g_{\ell-1}, h_\ell, g_\ell, \ldots, g_{m-1}) = \eta \cdot \lambda^i(h_\ell).$$

For a contradiction, let us assume $g \triangleq h$. Then applying (1c) with

$$\xi = (\lambda_0, \ldots, \lambda_{\ell-1}, \lambda_\ell, \lambda_\ell, \lambda_{\ell+1}, \ldots, \lambda_{m-1}),$$

$$\xi' = (\lambda_0, \ldots, \lambda_{\ell-1}, h^*, \lambda_\ell, \lambda_{\ell+1}, \ldots, \lambda_{m-1}) \in [A]\binom{m+1}{m}),$$

where $h^* = h_\ell$ if $h_\ell \in A$ and $h^* = \lambda_{\min h^{-1}(\lambda_j)}$ if $h_\ell = \lambda_j$, we get

$$\xi \cdot g = (g_0, \ldots, g_{\ell-1}, g_\ell, g_\ell, \ldots, g_{m-1})$$

$$\triangleq (h_0, \ldots, h_{\ell-1}, h_\ell, h_\ell, \ldots, h_{m-1}) = \xi \cdot h,$$

and

$$\xi' \cdot g = (g_0, \ldots, g_{\ell-1}, h_\ell, g_\ell, \ldots, g_{m-1})$$

$$\overset{\triangle}{=} (h_0, \ldots, h_{\ell-1}, h_\ell, h_\ell, \ldots, h_{m-1}) = \xi' \cdot h.$$

Note that $\xi' \cdot g = (g_0, \ldots, g_{\ell-1}, h_\ell, g_\ell, \ldots, g_{m-1})$ comes from the fact that $\min g^{-1}(\lambda_j) = \min h^{-1}(\lambda_j)$, in case $h_\ell = \lambda_j$. Therefore, by transitivity we have

$$(g_0, \ldots, g_{\ell-1}, g_\ell, g_\ell, \ldots, g_{m-1}) \overset{\triangle}{=} (g_0, \ldots, g_{\ell-1}, h_\ell, g_\ell, \ldots, g_{m-1}),$$

which contradicts (6.3). Hence $g \overset{\triangle}{\neq} h$, which completes the proof of Theorem 6.6.

□

This result was proved in Prömel and Voigt (1983), cf. also Prömel and Voigt (1986).

6.4 Applications

Every result which can be proved using the Graham-Rothschild theorem admits some kind of canonization using the canonizing version of Graham-Rothschild's theorem instead. Here we will only discuss three examples where applying the canonizing Graham-Rothschild theorem easily gives a canonical set of equivalence relations.

6.4.1 Finite Unions and Finite Sums

The first application of the canonical Graham-Rothschild theorem is a canonizing version of the finite union theorem (cf. Sect. 5.2.4). Recall that every nonempty subset of n can be interpreted as an element of $[1]\binom{n}{1}$. Observing that there are precisely three $(\{0\}, 1)$-canonical pairs, viz. $(\emptyset, (\{0, \lambda\})_{\approx_1}), (\{0\}, (\{0\}, \{\lambda\})_{\approx_1})$ and $(\{0\}, (\{0, \lambda\})_{\approx_1})$, we obtain

Theorem 6.7. *Let m be a positive integer. Then there exists $n = n(m)$ such that for every coloring $\Delta : \mathcal{B}(n) \to \omega$ there exist m mutually disjoint and non empty subsets $X_0, \ldots, X_{m-1} \in \mathcal{B}(n)$ such that one of the following three cases is valid for all nonempty $I, J \subseteq m$:*

(1) $\Delta(\bigcup_{i \in I} X_i) = \Delta(\bigcup_{j \in J} X_j)$
(2) $\Delta(\bigcup_{i \in I} X_i) = \Delta(\bigcup_{j \in J} X_j)$ *if and only if* $I = J$
(3) $\Delta(\bigcup_{i \in I} X_i) = \Delta(\bigcup_{j \in J} X_j)$ *if and only if* $\min I = \min J$.   □

Using again the bijection $\kappa : \mathcal{B}(n) \to 2^n$ given by $\kappa(B) = \sum_{i \in B} 2^i$ for every $B \subseteq n$ we obtain a canonical Rado-Folkman-Sanders theorem, viz.

Theorem 6.8. *Let m be a positive integer. Then there exists $n = n(m)$ such that for every coloring $\Delta : n \to \omega$ there exist mutually distinct positive integers $a_0, \ldots, a_{m-1}$ such that one of the following three cases is valid for all nonempty $I, J \subseteq m$:*

(1) $\Delta(\sum_{i \in I} a_i) = \Delta(\sum_{j \in J} a_j)$
(2) $\Delta(\sum_{i \in I} a_i) = \Delta(\sum_{j \in J} a_j)$ *if and only if* $I = J$
(3) $\Delta(\sum_{i \in I} a_i) = \Delta(\sum_{j \in J} a_j)$ *if and only if* $\min I = \min J$ $\square$

It is interesting to note that if finite subsets of ω are partitioned, instead of subsets of some finite n, respectively ω instead of n, and we ask for the canonical patterns on finite unions, respectively finite sums, then it turns out that three patterns are no longer sufficient (Taylor 1976).

6.4.2 Boolean Lattices

From the canonical Hales-Jewett theorem we obtained that coloring the points (i.e., $B(0)$-sublattices) of a sufficiently large Boolean lattice always yields a $B(m)$-sublattice which is either colored monochromatically or one-to-one (Corollary 6.2). Clearly, these two patterns do not longer suffice if we color $B(1)$-sublattices, i.e., 2-element chains.

Every 2-element chain in a Boolean lattice is given by a pair $(X_0, X_0 \cup X_1)$, where $X_1 \neq \emptyset$ and $X_0 \cap X_1 = \emptyset$. On the other hand every such chain can be interpreted as a one parameter word over the alphabet $\{0, 1\}$. Using this interpretation, the canonizing Graham-Rothschild theorem gives a canonical set of equivalence relations as follows: on the left hand side as $(2, 1)$-canonical pairs, on the right hand side in terms of 2-element chains saying that $(X_0, X_0 \cup X_1)$ is equivalent to $(Y_0, Y_0 \cup Y_1)$ if and only if the equation(s) in the second column is (are) fulfilled:

$J = \emptyset$ and

$\quad (\{0, \lambda\}, \{1\})_{\approx_1}$ $X_0 = Y_0$
$\quad (\{0\}, \{1, \lambda\})_{\approx_1}$ $X_0 \cup X_1 = Y_0 \cup Y_1$
$\quad (\{0, 1, \lambda\})_{\approx_1}$ always

$J = \{0\}$ and

$\quad (\{0\}, \{1\})_{\approx_0}, (\{0\}, \{1\}, \{\lambda\})_{\approx_1}$ $X_0 = Y_0$ and $X_1 = Y_1$
$\quad (\{0\}, \{1\})_{\approx_0}, (\{0, 1\}, \{\lambda\})_{\approx_1}$ $\{x \in X_0 \mid x < \min X_1\} = \{y \in Y_0 \mid y < \min Y_1\}$
$\qquad\qquad$ and $X_1 = Y_1$

$\quad (\{0\}, \{1\})_{\approx_0}, (\{0, \lambda\}, \{1\})_{\approx_1}$ $X_0 = Y_0$ and $\min X_1 = \min Y_1$
$\quad (\{0\}, \{1\})_{\approx_0}, (\{0\}, \{1, \lambda\})_{\approx_1}$ $X_0 \cup X_1 = Y_0 \cup Y_1$ and $\min X_1 = \min Y_1$
$\quad (\{0\}, \{1\})_{\approx_0}, (\{0, 1, \lambda\})_{\approx_1}$ $\{x \in X_0 \mid x < \min X_1\} = \{y \in Y_0 \mid y < \min Y_1\}$
$\qquad\qquad$ and $\min X_1 = \min Y_1$

$\quad (\{0, 1\})_{\approx_0}, \quad (\{0, 1\}, \{\lambda\})_{\approx_1}$ $X_1 = Y_1$
$\quad (\{0, 1\})_{\approx_0}, \quad (\{0, 1, \lambda\})_{\approx_1}$ $\min X_1 = \min Y_1$

In general, coloring $B(k)$-lattices one obtains a canonizing version in the same way interpreting the $(2, k)$-canonical pairs in terms of sets. For sublattices of

Boolean lattices, i.e., for arbitrary distributive lattices, the situation gets slightly more complicated. The interested reader will find a discussion of this in Prömel and Voigt (1982).

6.4.3 Finite Sets

The last application of the canonical Graham-Rothschild theorem we mention in this section is another proof of a finite version of the Erdős-Rado canonization theorem.

Theorem 6.9. *Let k and m be positive integers. Then there exists a positive integer $n = ER(k, m)$ such that for every coloring $\Delta : [n]^k \to \omega$ there exists an m-subset $M \in [n]^m$ and there exists a (possible empty) set $J \subseteq k$ such that*

$$\Delta(X) = \Delta(Y) \quad \text{if and only if} \quad X : J = Y : J$$

for every pair $X, Y \in [M]^k$.

Proof. Let n be according to Theorem 6.6 with respect to $A = \{0\}$, k and m. Let $\Delta : [n]^k \to \omega$ be a coloring. Define $\Delta' : [\{0\}]\binom{n}{k} \to \omega$ by $\Delta'(g) = \Delta(\phi \cdot g)$. Then there exist a $(\{0\}, k)$-canonical pair $\Pi = (J, (\approx_i)_{i \in J+})$ and an $f \in [\{0\}]\binom{n}{m}$ satisfying Theorem 6.6. Observe that by definition of Δ', every $\approx_i$ can only have one equivalence class. But this implies immediately that $M = \{f^{-1}(\lambda_i) \mid i < m\} \in [n]^m$ and $J \subseteq k$ satisfy Theorem 6.9. $\qquad\qquad\square$

Part III
Back to the Roots: Sets

Chapter 7
Ramsey Numbers

The finite version of Ramsey's theorem asserts that for every triple k, m and r of positive integers there exists a positive integer n such that $n \to (m)_r^k$. In Sect. 1.2 a compactness argument was used to derive this result from the infinite Ramsey theorem not giving any information about the size of n.

A lot of effort was spent during the last decades to get some information on the size of the least $n = RAM(k, m, r)$ satisfying Ramsey's theorem. Ramsey himself gave a constructive proof for the existence of n getting for example $m!$ as an upper bound for $RAM(2, m, 2)$, i.e., for the least n such that $n \to (m)_2^2$. But he already admitted that "this value is, I think, still much too high" and gave some advices to lower this bound. A few years later Erdős and Szekeres (1935) obtained a new proof of Ramsey's theorem yielding a better upper bound for $RAM(2, m, 2)$, viz. $O(4^m m^{-1/2})$, cf. Sect. 7.1. This is essentially still the best known value. On the other hand, the best lower bound obtained so far is $\Omega(m 2^{m/2})$ leaving quite a big gap for the actual growth of the function $RAM(2, m, 2)$. This lower bound was proved by Spencer (1975a) using probabilistic means, cf. Sect. 7.3. For general values of k and r the situation is even more distressing, as it will be seen in Sect. 7.5. In Sect. 7.4 we consider the so called off-diagonal Ramsey numbers giving the least value $R(s, t)$ such that for every 2-coloring of the pairs in $R(s, t)$ there exists either a monochromatic s-subset in color 0 or a monochromatic t-subset in color 1. Here for $s = 3$ the known lower and the upper bounds match asymptotically, but for $s \geq 4$ the known bounds are till far apart.

7.1 The Finite Ramsey Theorem: A Constructive Proof

First observe that it is enough to prove the finite Ramsey theorem for two colors. Assume that for some $r \geq 2$ we know for every pair k, m the existence of an $n = n(k, m, r)$ so that $n \to (m)_r^k$. Then $n' = n(k, n(k, m, 2), r)$ fulfills the theorem for k, m and $r + 1$, as the following argument shows.

H.J. Prömel, *Ramsey Theory for Discrete Structures*,
DOI 10.1007/978-3-319-01315-2_7,
© Springer International Publishing Switzerland 2013

Let an arbitrary $r + 1$-coloring of $[n']^k$ be given. Then consider this as an r-coloring of $[n']^k$ simply by combining the colors $r - 1$ and r. By choice of n' there exists either a monochromatic $n(k, m, 2)$-subset of n' in one of the colors 0, ..., $r - 2$. In this case we are done. Or there exists an $n(k, m, 2)$-subset with all its k-subsets in color $r - 1$ or in color r. Then by choice of $n(k, m, 2)$ we find a monochromatic m-subset.

Let $R(k; s, t)$ denote the least n such that for every 2-coloring $\Delta : [n]^k \to 2$ there exists either an s-subset $S \in [n]^s$ such that $\Delta \rceil S^k$ is identically 0 or there exists a t-subset $T \in [n]^t$ so that $\Delta \rceil T^k$ is identically 1. For a more picturesque formulation one may freely replace the color 0 by 'red' and the color 1 by 'blue'.

To prove the finite Ramsey theorem we show

Theorem 7.1. *Let k, $s \geq k$ and $t \geq k$ be positive integers. Then*

$$R(k; s, t) \leq R\big(k - 1;\ R(k; s - 1, t),\ R(k; s, t - 1)\big) + 1.$$

Proof. The proof is given by induction on k and on s, t. Observe that, by the pigeonhole principle, $R(1; x, y) = x + y - 1$ for all x and y and, moreover, $R(\ell; x, \ell) = R(\ell; \ell, x) = x$ for all ℓ and x.

Now assume the existence of $R(k - 1; x, y)$ for all x and y and of $R(k; s - 1, t)$ and $R(k; s, t - 1)$.

Put $n = R(k - 1;\ R(k; s - 1, t),\ R(k; s, t - 1)) + 1$ and assume an arbitrary 2-coloring of $[n]^k$ to be given. This induces a 2-coloring on $[n - 1]^{k-1}$ by coloring $X \in [n - 1]^{k-1}$ in the same color as $X \cup \{n - 1\}$. By choice of n and by symmetry we can assume to find an $R(k; s - 1, t)$-subset $Y \subseteq n - 1$ which is monochromatic in color 0.

Now consider the coloring of $[Y]^k$. According to its size Y contains either a monochromatic subset of size t in color 1. Then we are done. Or Y contains a monochromatic subset Z of size $s - 1$ in color 0. Then $Z \cup \{n - 1\}$ is the desired s-subset of n in color 0.   □

Upper bounds for the Ramsey-function in general are given in Erdős and Rado (1952). We consider only the case $k = 2$ explicitly. Observe that this case can be interpreted in terms of graphs by saying that every graph with at least $R(2; s, t)$ vertices contains either a stable set on s vertices or a clique on t vertices.

Corollary 7.2. *Let $s, t \geq 2$. Then*

$$R(2; s, t) \leq \binom{s + t - 2}{s - 1}.$$

Proof. Theorem 7.1 and the pigeonhole principle give that

$$R(2; s, t) \leq R(1; \ R(2; s-1, t), \ R(2; s, t-1)) + 1$$

$$= R(2; s-1, t) + R(2; s, t-1).$$

Moreover recall that $R(2; s, 2) = s$ and $R(2; 2, t) = t$ for every $s, t \geq 2$. The claim follows then easily by induction.   □

7.2 Some Exact Values

Much effort has been spent on determining exact values of the Ramsey function with distressingly little success even in case $k = 2$. We abbreviate $R(2; s, t)$ by $R(s, t)$.

It is very easy to see that $R(3, 3) = 6$. Obviously, $R(3, 3) \leq 6$ and the pentagon C_5 which neither contains a stable set on 3 vertices nor a triangle shows that $R(3, 3) > 5$. It is also not difficult to determine $R(4, 4)$.

Theorem 7.3.

$$R(4, 4) = 18.$$

Proof. Notice that $R(s, t) = R(t, s)$ for all s, t. From Theorem 7.1 it follows that $R(4, 4) \leq 2 \cdot R(3, 4)$. We show that $R(3, 4) \leq 9$. Assume that $[9]^2$ is 2-colored without a 3-subset in color 0 and a 4-subset in color 1. Then each element $x < 9$ is contained in precisely three 2-subsets colored with 0 and five 2-subsets colored with 1. To see this recall that $R(2, 4) = 4$ and $R(3, 3) = 6$. Thus, there are exactly $\frac{9 \cdot 3}{2}$ many 2-subsets colored with 0. But this should be an integer. A contradiction! So we have that $R(3, 4) \leq 9$ and therefore $R(4, 4) \leq 18$. The graph in Fig. 7.1 shows that equality holds. The vertices are $\mathbb{Z}_{17}$ and $\{i, j\}$ is an edge if and only if $i - j$ is a square in $\mathbb{Z}_{17}$.   □

Despite all efforts these are the only diagonal Ramsey numbers known! Both have been found already in 1955 by Greenwood and Gleason, as well as the off-diagonal Ramsey numbers $R(3, 4) = 9$ and $R(3, 5) = 14$.

The difficulty in obtaining diagonal Ramsey numbers for larger values is perhaps best illustrated by a quote from Erdős (1985): "Suppose an evil spirit would tell us, 'Unless you tell me the value of $R(5, 5)$ I will exterminate the human race.' Our best strategy would perhaps be to get all the computers and computer scientists to work on it. If he would ask for $R(6, 6)$ our best bet would perhaps be to try the destroy him before he destroys us".

We collect all known values as well as the best known bounds for $R(s, t), s, t \leq 7$ in Fig. 7.2. The Ramsey number $R(3, 6) = 18$ is due to Kalbfleisch (1966) and $R(3, 7) = 23$ is from Graver and Yackel (1968). The values in Fig. 7.2 are from the dynamic survey by Radziszowski (2011) that also contains the remaining references.

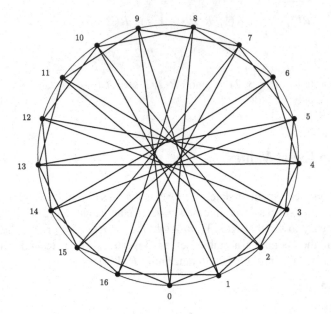

Fig. 7.1 $R(4, 4) > 17$

$s\backslash t$	3	4	5	6	7
3	6	9	14	18	23
4		18	25	35–41	49–61
5			43–49	58–87	80–143
6				102–165	113–298
7					205–540

Fig. 7.2 Some known exact values and bounds for $R(s, t)$

7.3   A Lower Bound for Diagonal Ramsey Numbers

By Corollary 7.2 we have that

$$R(s, s) \leq \binom{2s-2}{s-1} = (\tfrac{1}{4\sqrt{\pi}} + o(1))4^s s^{-1/2},$$

hence an exponential upper bound for $R(s, s)$. Erdős (1947) was the first to prove an exponential lower bound for $R(s, s)$. This was one of the earliest applications of the nonconstructive (probabilistic) method in combinatorics.

Theorem 7.4.

$$R(s, s) > \tfrac{s}{e\sqrt{2}} \cdot 2^{s/2}.$$

Proof. We show that $\binom{n}{s}2^{1-\binom{s}{2}} < 1$ implies that $R(s, s) > n$. Let $\mathbb{A}_n \subseteq [n]^2$ be a random subset. More precisely, $\mathbb{A}_n$ is a random variable with values being subsets of $[n]^2$. We require that

$$\text{Prob}[\{i, j\} \in \mathbb{A}_n] = \tfrac{1}{2} \quad \text{for every } \{i, j\} \in [n]^2$$

and that all these events are mutually independent. Hence, $\mathbb{A}_n$ can be viewed as the edge set of a random graph on the vertex set n with edge probability $1/2$.

For $S \in [n]^s$ let $\mathbb{A}_S$ be the event that S forms either a complete subgraph or a stable set, in other words $[S]^2 \subseteq \mathbb{A}_n$ or $[S]^2 \cap \mathbb{A}_n \neq \emptyset$. Then

$$\text{Prob}[\mathbb{A}_S] = 2^{1-\binom{s}{2}}.$$

So the probability of the event that some s-element subset of n forms a clique or a stable set is given by

$$\text{Prob}\Big[\bigcup_{\substack{S \subseteq n \\ |S|=s}} \mathbb{A}_s\Big] \leq \sum_{\substack{S \subseteq n \\ |S|=s}} \text{Prob}[\mathbb{A}_s] \leq \binom{n}{s} \cdot 2^{1-\binom{s}{2}}.$$

For $\binom{n}{s} \cdot 2^{1-\binom{s}{2}} < 1$ this gives the existence of a graph on n having neither an s-clique nor a stable set on s vertices and therefore $R(s, s) > n$. It remains to calculate that $n = \frac{s}{e\sqrt{2}}2^{s/2}$ implies $\binom{n}{s}2^{1-\binom{s}{2}} < 1$. We leave this to the reader. $\quad\square$

Theorem 7.4 can be slightly improved, increasing the lower bound by a factor of 2. For such an improvement we need a very powerful tool, the so-called Lovász Local Lemma. The theorem we are going to prove can be found in Spencer (1977). This result is stronger than needed for bounding $R(s, s)$, but its generality will be of use later.

Let Ω be a probability space and $\mathbb{A}_0, \ldots, \mathbb{A}_{n-1}$ be n events. A graph $G = (n, E)$ is a dependence graph of $(\mathbb{A}_i)_{i<n}$ if for every $i < n$ the event $\mathbb{A}_i$ is mutually independent of $\{\mathbb{A}_j \mid \{i, j\} \notin E\}$. Note: The requirement that $\mathbb{A}_i$ is not only independent of each $\mathbb{A}_j$ but of any combination of the $\mathbb{A}_j$ is essential.

Theorem 7.5 (Lovász Local Lemma).   *Let $\mathbb{A}_0, \ldots, \mathbb{A}_{n-1}$ be events in a probability space Ω and let $G = (n, E)$ be a dependence graph of $(\mathbb{A}_i)_{i<n}$. Suppose there are $x_0, \ldots, x_{n-1}$ so that $0 < x_i < 1$ and*

$$\text{Prob}[\mathbb{A}_i] \leq x_i \cdot \prod_{\{i,j\}\in E} (1 - x_j) \quad \text{for every } i < n. \tag{7.1}$$

Then $\text{Prob}[\bigcap_{j<n} \overline{\mathbb{A}}_j] > 0$, where $\overline{\mathbb{A}}_j = \Omega \setminus \mathbb{A}_j$ is the complementary event.

Proof. We show by induction on $|J|$ that for every $J \subseteq n$ and every $i \notin J$ we have that

$$\text{Prob}[\mathbb{A}_i \mid \bigcap_{j \in J} \overline{\mathbb{A}}_j] \leq x_i. \tag{7.2}$$

Then this implies that

$$\begin{aligned}
\text{Prob}[\bigcap_{j<n} \overline{\mathbb{A}}_j] &= \text{Prob}[\overline{\mathbb{A}}_0 \mid \bigcap_{j \in [1,n-1]} \overline{\mathbb{A}}_j] \cdot \text{Prob}[\bigcap_{j \in [1,n-1]} \overline{\mathbb{A}}_j] \\
&\geq (1 - x_0) \cdot \text{Prob}[\bigcap_{j \in [1,n-1]} \overline{\mathbb{A}}_j] \\
&\geq \prod_{j<n}(1 - x_j) > 0,
\end{aligned}$$

proving the theorem.

By (7.1) it follows in particular that the assertion (7.2) is true for $J = \emptyset$. Fix $J \subseteq n$ with $|J| \geq 1$ and $i \notin J$. Let $I = \{j < n \mid \{i, j\} \in E\}$. If $I \cap J = \emptyset$ the claim holds vacuously. So assume that $I \cap J \neq \emptyset$. By relabeling the events $\mathbb{A}_j$ we may assume without loss of generality that the intersection $I \cap J$ consists of the first k events, i.e. $I \cap J = k$. Then

$$\begin{aligned}
\text{Prob}[\mathbb{A}_i \mid \bigcap_{j \in J} \overline{\mathbb{A}}_j] &= \text{Prob}[\mathbb{A}_i \mid \bigcap_{j<k} \overline{\mathbb{A}}_j \cap \bigcap_{j' \in J \backslash k} \overline{\mathbb{A}}_{j'}] \\
&= \frac{\text{Prob}[\mathbb{A}_i \cap \bigcap_{j<k} \overline{\mathbb{A}}_j \mid \bigcap_{j' \in J \backslash k} \overline{\mathbb{A}}_{j'}]}{\text{Prob}[\bigcap_{j<k} \overline{\mathbb{A}}_j \mid \bigcap_{j' \in J \backslash k} \overline{\mathbb{A}}_{j'}]}.
\end{aligned}$$

The denominator can be bounded as follows:

$$\text{Prob}[\bigcap_{j<k} \overline{\mathbb{A}}_j \mid \bigcap_{j' \in J \backslash k} \overline{\mathbb{A}}_{j'}] = \prod_{j<k} \text{Prob}[\overline{\mathbb{A}}_j \mid \bigcap_{j''=j+1}^{k-1} \overline{\mathbb{A}}_{j''} \cap \bigcap_{j' \in J \backslash k} \overline{\mathbb{A}}_{j'}] > \prod_{j<k}(1 - x_j),$$

using the inductive hypothesis. The numerator is estimated from above by

$$\text{Prob}[\mathbb{A}_i \cap \bigcap_{j<k} \overline{\mathbb{A}}_j \mid \bigcap_{j' \in J \backslash k} \overline{\mathbb{A}}_{j'}] \leq \text{Prob}[\mathbb{A}_i \mid \bigcap_{j \in J \backslash k} \overline{\mathbb{A}}_{j'}] \leq \text{Prob}[\mathbb{A}_i],$$

where the last equality holds since $\mathbb{A}_i$ is mutually independent of $\{\mathbb{A}_{j'} \mid j' \in J \backslash k\}$. Hence we get

$$\text{Prob}[\mathbb{A}_i \mid \bigcap_{j \in J} \overline{\mathbb{A}}_j] \leq \frac{\text{Prob}[\mathbb{A}_i]}{\prod_{j<k}(1 - x_j)} \overset{(7.1)}{\leq} \frac{x_i \cdot \prod_{j \in I}(1 - x_j)}{\prod_{j<k}(1 - x_j)} \overset{k \subseteq I}{\leq} x_i,$$

completing the proof of Theorem 7.5.   □

Combining the Lovász Local Lemma and the proof of Theorem 7.4 one obtains easily:

Theorem 7.6.

$$R(s,s) > (\tfrac{\sqrt{2}}{e} - o(1)) \cdot s \cdot 2^{s/2}.$$

Proof. We show that $e\binom{s}{2}\binom{n}{s-2}2^{1-\binom{s}{2}} < 1$ implies that $R(s,s) > n$.

Let the random subset $\mathbb{A}_n \subseteq [n]^2$ as well as the events $\mathbb{A}_S$ for $S \in [n]^s$ be as in the proof of Theorem 7.4. Recall that

$$\text{Prob}[\mathbb{A}_S] = 2^{1-\binom{s}{2}}.$$

Moreover, observe that $\mathbb{A}_S$ is independent of $\mathbb{A}_T$ provided that $|S \cap T| \leq 1$. Hence, the dependence graph G of the events $\{\mathbb{A}_S \mid S \in [n]^s\}$ has maximal degree less than $d = \binom{s}{2}\binom{n}{s-2} - 1$.

Let $x_0 = \ldots = x_{n-1} = \frac{1}{d+1}$. Then condition (7.1) in Theorem 7.4 becomes

$$\text{Prob}[\mathbb{A}_S] \leq \tfrac{1}{d+1}\left(\tfrac{d}{d+1}\right)^d. \tag{7.3}$$

Notice that $\frac{1}{e} < (\frac{d}{d+1})^d$. So, to verify (7.3) it is enough to observe that

$$\text{Prob}[\mathbb{A}_S] \cdot (d+1) \cdot e < 1,$$

or that

$$e\binom{s}{2}\binom{n}{s-2}2^{1-\binom{s}{2}} < 1.$$

Hence, assuming that $e\binom{s}{2}\binom{n}{s-2}2^{1-\binom{s}{2}} < 1$ allows to apply the Lovász Local Lemma in order to deduce that $R(s,s) > n$.

The calculation that $n = (\frac{\sqrt{2}}{e} - o(1))s2^{s/2}$ implies that $\binom{s}{2}\binom{n}{s-2}2^{1-\binom{s}{2}} < 1$ is left to the reader.   □

Collecting the upper and lower bound for $R(s,s)$ we obtain

$$\sqrt{2} \leq \liminf(R(s,s))^{1/s} \leq \limsup(R(s,s))^{1/s} \leq 4.$$

Surprisingly, even the existence of $\lim(R(s,s))^{1/s}$ has not yet been proved. This, together with determining its value (provided it exists), is one of the major open problems in Ramsey theory. P. Erdős did put quite a high amount of money on its solution (cf. Erdős 1981).

The lower bound given in Theorem 7.6 due to Spencer (1975a) is the best one known. The best upper bound is due to Conlon (2009) who improved the bound $R(s+1, s+1) \leq \binom{2s}{s}$ from Corollary 7.2 to

$$R(s+1, s+1) \leq s^{-c \frac{\log s}{\log \log s}} \binom{2s}{s},$$

for an appropriately chosen constant $c > 0$.

7.4   Asymptotics for Off-Diagonal Ramsey Numbers

From the proof of the finite Ramsey theorem, i.e., from Corollary 7.2 we get that for every fixed $s \geq 3$

$$R(s, t) \leq c_s t^{s-1},$$

where the constant c_s depends only on s. The first lower bound on $R(3, t)$ was proved by Erdős (1961) who showed that $R(3, t) \geq ct^2 (\log t)^{-2}$ for an appropriate constant c. Lower bounds for $R(s, t)$ in general were first given by Spencer (1975a). Here we present the slightly improved bound from Spencer (1977), cf. also Bollobás (2001).

Theorem 7.7. *Let $s \geq 3$ be a positive integer. Then there exists a constant $c = c(s)$ such that*

$$R(s, t) \geq c \cdot \left(\frac{t}{\ln t} \right)^{\frac{s+1}{2}}.$$

Proof. Let $\mathbb{A}_{n,p} \subseteq [n]^2$ be a random subset such that $\mathrm{Prob}[\{i, j\} \in \mathbb{A}_{n,p}] = p$ for every $\{i, j\} \in [n]^2$ and that all these events are mutually independent. As in previous proofs, $\mathbb{A}_{n,p}$ is viewed as a random graph. For $S \in [n]^s$, $T \in [n]^t$ let $\mathbb{A}_S$ be the event that S forms a clique, meaning that $[S]^2 \subseteq \mathbb{A}_{n,p}$, and let $\mathbb{B}_T$ be the event that T is a stable set, i.e. $[T]^2 \cap \mathbb{A}_{n,p} = \emptyset$. Then

$$\mathrm{Prob}[\mathbb{A}_S] = p^{\binom{s}{2}} =: p_0 \quad \text{for every } S \in [n]^s$$

and

$$\mathrm{Prob}[\mathbb{B}_T] = (1 - p)^{\binom{t}{2}} =: p_1 \quad \text{for every } T \in [n]^t.$$

By definition the inequality

$$\text{Prob}\Big[\bigcap_{S\in[n]^s} \overline{\mathbb{A}_S} \cap \bigcap_{T\in[n]^t} \overline{\mathbb{B}_T}\Big] > 0 \tag{7.4}$$

implies the existence of a graph without a cliques of size s and without any stable set of size t, showing that $R(3,t) > n$.

We use the Lovász Local Lemma to show that there exists a $c = c(s) > 0$ such that for every $n \leq c \cdot (t/\ln t)^{(s+1)/2}$ the inequality (7.4) is fulfilled for all t larger than some t_0. In order to show this we assume in the following that s is fixed, while t tends to infinity.

Let G be the dependence graph of the events $\mathbb{A}_S$, $S \in [n]^s$, and $\mathbb{B}_T$, $T \in [n]^t$, with vertex set $V(G) = [n]^s \cup [n]^t$. The edge set $E(G)$ is defined as follows: two sets are connected by an edge if and only if they intersect in at least two vertices.

Recall that for graphs $G = (V, E)$ and vertices $x \in V$, the set of neighbors of x is denoted by $\Gamma(x)$, i.e., $\Gamma(x) = \{y \in V \mid \{x, y\} \in E\}$. So we have for every $S \in [n]^s$ and $T \in [n]^t$ that

$$|\Gamma(S) \cap [n]^s| \leq \binom{s}{2}\binom{n}{s-2} \leq s^2 n^{s-2} =: d_{00},$$

$$|\Gamma(S) \cap [n]^t| \leq \binom{s}{2}\binom{n}{t-2} \leq n^{t-2} \quad =: d_{01},$$

and

$$|\Gamma(T) \cap [n]^s| \leq \binom{t}{2}\binom{n}{s-2} \leq t^2 n^{s-2} =: d_{10},$$

$$|\Gamma(T) \cap [n]^t| \leq \binom{t}{2}\binom{n}{t-2} \leq n^{t-2} \quad =: d_{11},$$

where in the second and forth line we used our assumption that t is large. To apply the Lovász Local Lemma we have to show the existence of positive numbers x_S, $S \in [n]^s$ and x_T, $T \in [n]^t$, all smaller than 1, such that for every $S \in [n]^s$

$$p_0 = \text{Prob}[\mathbb{A}_s] \leq x_S \cdot \prod_{S'\in\Gamma(S)\cap[n]^s} (1 - x_{S'}) \cdot \prod_{T\in\Gamma(S)\cap[n]^t} (1 - x_T)$$

and for every $T \in [n]^T$

$$p_1 = \text{Prob}[\mathbb{B}_T] \leq x_T \cdot \prod_{S\in\Gamma(T)\cap[n]^s} (1 - x_S) \cdot \prod_{T'\in\Gamma(T)\cap[n]^t} (1 - x_{T'}).$$

Let $x_S = x_0$ for every S and $x_T = x_1$ for every T. Observe that for $x < 0.68$ we have $\ln(1 - x) > -x(1 + x)$. Therefore, it is enough to show that there exist $x_0, x_1 < 0.68$ such that

$$\ln \frac{x_0}{p_0} > d_{00}x_0(1 + x_0) + d_{01}x_1(1 + x_1), \tag{7.5}$$

$$\ln \frac{x_1}{p_1} > d_{10}x_0(1 + x_0) + d_{11}x_1(1 + x_1). \tag{7.6}$$

Choose a positive constant $c < 1$ (as indicated below) and put $n = c \cdot (t / \ln t)^{(s+1)/2}$, $p = 12s \cdot \ln t / t$ and let

$$x_0 := 2p_0 = o(1)$$

$$x_1 := t^{ts} p_1 = t^{ts} \cdot (1 - p)^{\binom{t}{2}} < t^{ts} \cdot e^{-p\binom{t}{2}} = t^{ts} \cdot t^{-\frac{12s(t-1)}{2}} < t^{-2ts} = o(1).$$

To verify (7.5) and (7.6) observe first that

$$n^{s-2} p^{\binom{s}{2}} = (12s)^{\binom{s}{2}} c^{s-2} \left(\frac{t}{\ln t}\right)^{\frac{(s-2)(s+2)}{2}} \cdot \left(\frac{\ln t}{t}\right)^{\binom{s}{2}} = (12s)^{\binom{s}{2}} c^{s-2} \frac{\ln t}{t}$$

and that

$$n^{t-2} x_1 \leq \left(\frac{t}{\ln t}\right)^{\frac{(t-2)(s+1)}{2}} t^{-2s} < t^{ts} t^{-2ts} = t^{-ts},$$

for t sufficiently large. From these bounds we immediately get

$$d_{00}x_0 < 2s^2 (12s)^{\binom{s}{2}} c^{s-2} \frac{\ln t}{t} = o(1),$$

$$d_{01}x_1 = d_{11}x_1 < t^{-ts} = o(1),$$

$$d_{10}x_0 \leq 2t^2 (12s)^{\binom{s}{2}} c^{s-2} \cdot \frac{\ln t}{t} = 2(12s)^{\binom{s}{2}} c^{s-2} t \ln t.$$

From the definition of x_0 and x_1 we thus easily deduce that for t sufficiently large (7.5) and (7.6) are satisfied if $c = c(s) > 0$ is defined such that $s > 2(12s)^{\binom{s}{2}} c^{s-2}$. This completes the proof of Theorem 7.7.   □

Graver and Yackel (1968) were the first to improve the upper bound derived from the finite Ramsey theorem by showing that for every fixed s there is a constant c_s so that $R(s,t) \leq c_s t^{s-1} (\ln t)^{-1} \ln \ln t$. This yields in particular $R(3,t) \leq c_3 t^2 (\ln t)^{-1} \ln \ln t$. Ajtai et al. (1980, 1981) succeeded in getting rid of the $\ln \ln$-term by proving

Theorem 7.8.

$$R(3,t) \leq \frac{t^2}{\ln t - 1}.$$

What Ajtai, Komlós and Szemerédi actually did was proving a lower bound for the independence number $\alpha(G)$ of triangle free graphs G.

Let G be a graph on n, let $e = e(G)$ denote the number of edges of G and let $d = d(G)$ be the average degree in G, i.e., $d = \frac{2e(G)}{n}$.

Lemma 7.9. *For every triangle-free graph $G = (n, E)$ the following inequality is valid:*

$$\alpha(G) \geq n \cdot \frac{\ln d - 1}{d}.$$

We first indicate how Theorem 7.8 follows from Lemma 7.9.

Proof of Theorem 7.8. Assume there exists a triangle free graph G on $t^2/(\ln t - 1)$ vertices such that its independence number is at most $t - 1$. Since G is triangle free, the degree of every vertex in G is at most $t - 1$ and, therefore, $d(G) \leq t - 1$. Using the lemma we deduce

$$t - 1 \geq \alpha(G) \geq \frac{t^2}{\ln t - 1} \cdot \frac{\ln t - 1}{t} = t,$$

which is a contradiction.   □

Proof of Lemma 7.9. We follow an account given by Shearer (1983). The idea is to prove a slightly stronger result than stated in the lemma. Namely, $\alpha(G) \geq n\, f(d)$, where

$$f(d) = \frac{d \ln d - d + 1}{(d-1)^2} \quad \text{for } d \neq 0, 1 \quad \text{and } f(0) = 1, \ f(1) = \frac{1}{2} \qquad (7.7)$$

is a seemingly complicated function – chosen in such a way that the calculations below work out nicely.

Observe that $\alpha(G) \geq n\, f(d)$ implies the lemma, as one easily checks that $f(d) \geq (\ln d - 1)/d$. For the proof that $\alpha(G) \geq n\, f(d)$ we first note that f has some nice properties: f is continuous on $\mathbb{R}^+$, $f'(d) < 0$ and $f''(d) \geq 0$ and, moreover, f satisfies the differential equation

$$(d+1)f(d) = 1 + (d - d^2)f'(d), \qquad (7.8)$$

for all $d \in \mathbb{R}^+$.

We prove the claim by induction on n, the case $n = 0$ being trivial. So assume the claim is true for some $n \geq 0$. For a vertex $x \in V$ let $\deg(x)$ denote the degree of x and let $D_\Gamma(x)$ be the sums of the degrees of the neighbors of x, i.e.,

$$D_\Gamma(x) = \sum_{y \in \Gamma(x)} \deg(y).$$

Observe that

$$\sum_{x \in V} D_\Gamma(x) = \sum_{x \in V} \sum_{y \in \Gamma(x)} \deg(y) = \sum_{x \in V} (\deg(x))^2 \geq nd^2. \qquad (7.9)$$

For a vertex $x \in V$ let G_x be the induced subgraph on $V \backslash (\{x\} \cup \Gamma(x))$, i.e., the subgraph of G formed by deleting x and all its neighbors and let $d_x = d(G_x)$. Obviously,

$$|V(G_x)| = n - \deg(x) - 1$$

and, since G is triangle-free

$$|E(G_x)| = \frac{d}{2}n - D_\Gamma(x).$$

Therefore,

$$d_x = (dn - 2D_\Gamma(x))/(n - \deg(x) - 1).$$

Now applying the inductive hypothesis yields a stable set of size

$$(n - \deg(x) - 1) \cdot f(d_x)$$

in G_x. Adding x to this set gives an independent set of size

$$1 + (n - \deg(x) - 1) \cdot f(d_x)$$

in G. Taking the average over all vertices $x \in V$ we get a lower bound for $\alpha(G)$, i.e.,

$$\alpha(G) \geq 1 + \frac{1}{n} \sum_{x \in V} (n - \deg(x) - 1) \cdot f(d_x)$$

Recalling that $f''(d) \geq 0$ we have

$$f(d_x) \geq f(d) + (d_x - d) \cdot f'(d).$$

Hence, putting things together we get

$$\alpha(G) \geq 1 + \frac{1}{n} \sum_{x \in V} (n - \deg(x) - 1) \cdot (f(d) + (d_x - d) \cdot f'(d))$$

$$= 1 + (n - d - 1) \cdot (f(d) - df'(d)) + \frac{1}{n} \sum_{x \in V} (dn - 2D_\Gamma(x)) \cdot f'(d)$$

$$\overset{(7.9), \, f'<0}{\geq} 1 + (n - d - 1) \cdot (f(d) - df'(d)) + (dn - 2d^2) \cdot f'(d)$$

$$\overset{(7.8)}{=} nf(d).$$

Thus G contains a stable set of size $nf(d)$, thus completing the proof of Lemma 7.9.

$\square$

Ajtai et al. (1980) also showed that Lemma 7.9 can be used to prove inductively an upper bound for all off-diagonal Ramsey numbers.

Theorem 7.10. *There exists a constant $c > 0$ such that for all $s, t \geq 3$*

$$R(s, t) \leq \frac{(ct)^{s-1}}{(\ln t)^{s-2}}.$$

Proof. We proceed by induction on s. From Theorem 7.8 we know that the claim is true for $s = 3$ and all $t \geq 3$ for, say, $c = 12$. So assume now that there exists an $s \geq 3$ so that the claim holds for all $s' \leq s$. Fix some t and assume there exists a graph G on $n := \frac{(ct)^s}{(\ln t)^{s-1}}$ vertices without a clique on $s + 1$ vertices and with $\alpha(G) \leq t - 1$. Clearly, the degree of every vertex in G is at most $R(s, t)$, which by induction hypotheses is at most $\frac{(ct)^{s-1}}{(\ln t)^{s-2}} =: d$. Hence, the number of edges in G is bounded by $nd/2$. Let t_G denote the number of triangles in G. We proceed by a case distinction on the size of t_G. For this choose $\epsilon > 0$ so that $(s - 1)(1 - \epsilon) = s - 2$, i.e. $\epsilon = 1/(s - 1)$.

Assume first that $t_G \geq 3nd^{2-\epsilon}$. Then there exists a vertex v that is contained in at least $d^{2-\epsilon}$ triangles. Thus, the neighborhood $\Gamma(v)$ induces a subgraph G' on at most $n' = |\Gamma(v)| \leq d$ vertices that contains at least $d^{2-\epsilon}$ edges. G' thus contains a vertex w such that

$$|\Gamma(v) \cap \Gamma(w)| \geq d^{1-\epsilon} = \left(\frac{(ct)^{s-1}}{(\ln t)^{s-2}} \right)^{1-\epsilon} \geq \frac{(ct)^{s-2}}{(\ln t)^{s-3}} \overset{i.h.}{\geq} R(s - 1, t),$$

thus contradicting our assumptions on G.

So assume now that $t_G \leq 3nd^{2-\epsilon}$. We aim at showing that this implies that $\alpha(G) \geq t$, which again contradicts our assumption on G and will thus conclude the proof of Theorem 7.10. The idea is to choose an appropriate subset of vertices $V' \subseteq V(G)$ such that the induced graph $G[V']$ has no triangles, and then use Lemma 7.9 to deduce $\alpha(G[V']) \geq t$. To this end let $p := d^{-1+\frac{1}{4}\epsilon}$ and consider a random subset $V_p \subseteq G(V)$, such that each vertex of G is included with probability p, independently. Clearly, the size of the subset V_p is binomially distributed with probability p. Using, for example, Chebyshev inequality we deduce that $\text{Prob}[|V_p| < \frac{1}{2}np] < 1/4$. Now, consider a subgraph of G induced by the subset V_p and let E_p and T_p be the set of edges and triangles in $G[V_p]$. A simple calculation yields that the expected size of E_p is at most $\frac{nd}{2} \cdot p^2$, and the expected size of T_p is $t_G \cdot p^3$. Since $|E_p|$ and $|T_p|$ are nonnegative random variables, by Markov inequality we have $\text{Prob}[|E_p| > 4\frac{nd}{2}p^2] < 1/4$ and $\text{Prob}[|T_p| > 4t_G p^3] < 1/4$. From the union bound we thus get that there exists an induced subgraph G' of G on at least $\frac{1}{2}np = \frac{1}{2}nd^{-1+\epsilon/4}$ vertices, with at most $4\frac{nd}{2}p^2 = 2nd^{-1+\epsilon/2}$ edges and at most $4t_G p^3 = 12nd^{-1-\epsilon/4}$ triangles. Remove one vertex from each triangle. This leaves us with an induced subgraph G'' of G on, say, at least $\frac{1}{4}nd^{-1+\epsilon/4}$ vertices and with at most $2nd^{-1+\epsilon/2}$ edges (and thus average degree $\hat{d} \leq 16d^{\epsilon/4}$) that is triangle-free. From Lemma 7.9 we thus deduce that

$$\alpha(G) \geq \alpha(G'') \geq \frac{1}{64}\frac{n}{d} \cdot \frac{1}{4}\epsilon \ln d \geq \frac{\epsilon}{256}\frac{ct}{\ln t} \cdot ((s-1)\ln(ct) - (s-2)\ln\ln t).$$

Recalling that we did set $\epsilon = 1/(s-1)$ we thus deduce that $\alpha(G) \geq t$, for all $c \geq 500$, say. Thus concluding the proof of Theorem 7.10.   □

Comparing Theorems 7.7 and 7.8 shows that in case of $R(3, t)$ the upper and lower bounds are within a factor of $\ln t$. In a breakthrough paper Kim (1995) improved the lower bound to $cn^2/\ln t$, thus matching the upper bound up to a constant factor. An alternative proof was later given by Bohman (2009).

While the asymptotics for the case $s = 3$ is now settled, the case for $s \geq 4$ is still wide open. It is conjectured that in general for $s \geq 3$ we should have $R(s, t) = t^{s-1+o(1)}$ asymptotically in t (see Spencer (1975a, 1977)). But this conjecture is far from being settled even in case $s = 4$. Bohman and Keevash (2010) improved the lower bound from Theorem 7.7 by a logarithmic factor. However, this still leaves the best lower bound at $t^{(s+2)/2+o(1)}$ and thus far away from the conjectured truth.

7.5 More than Two Colors

For an arbitrary number of colors, say r, we define $R(s_0, \ldots, s_{r-1})$ to be the least integer such that for every r-coloring of the pairs in $R = R(s_0, \ldots, s_{r-1})$ there exist a color $i < r$ and an s_i-subset $S_i \in [R]^i$ so that all pairs S_i have color i. A straightforward extension of the argument given in the proof of the finite Ramsey theorem yields

$$R(s_0, \ldots, s_{r-1}) \leq 2 + \sum_{i<r}[R(s_0, \ldots, s_i - 1, \ldots, s_{r-1}) - 1], \qquad (7.10)$$

as every vertex can be incident to at most $R(s_0, \ldots, s_i - 1, \ldots, s_{r-1}) - 1$ edges in color i without inducing one of the desired monochromatic subsets.

The situation for r-color Ramsey numbers for $r \geq 3$ is even more distressing than for the 2-color Ramsey numbers. The only known exact value is $R(3, 3, 3) = 17$ determined already in 1955 by Greenwood and Gleason. Since $R(3, 3, 2) = R(3, 3) = 6$ we get that $R(3, 3, 3) \leq 17$ immediately from (7.10). Now consider the elements of $GF(2^4)$ as the elements of the underlying set. Color a pair x, y with color 0 if the difference $x - y$ is a cubic residue, and with color i ($i = 1, 2$) if $x - y$ belongs to the ith coset of cubic residues in the multiplicative group of field elements. Notice that $-1 \equiv 1 \pmod 2$, hence the order of differencing does not matter. The so colored set does not contain a monochromatic triangle, thus $R(3, 3, 3) \geq 17$.

Let $R_r(3) = R(3, \ldots, 3)$ (r-times). Combining (7.10) and the fact that $R(3, 3) = 6$ one gets that

$$R_r(3) \leq 3(r!).$$

The trivial lower bound $2^r \leq R_r(3)$ was improved by Chung (1973) showing that

$$R_r(3) \geq 3R_{r-1}(3) + R_{r-3}(3) - 3.$$

Using that $R_3(3) = 17$ we get from this inequality that $R_4(3) \geq 51$. The best known upper bound for $R_4(3)$ is due to Fettes et al. (2004), viz. $R_4(3) \leq 62$.

Another way for obtaining lower bounds for $R_r(3)$ is to consider sum-free partitions of integers. A set $A \subseteq \mathbb{N}$ of positive integers is said to be *sum-free* if $a_i + a_j \notin A$ whenever $a_i, a_j \in A$. Let s_r be the largest integer such that $1, \ldots, s_r$ can be partitioned into r sum-free sets $A_1, \ldots, A_r$.

Now color the pair $\{a, b\}$, where $1 \leq a < b \leq s_f + 1$, with color i if $b - a \in A_i$. This coloring obviously does not produce any monochromatic triangle. Therefore one has that $R_r(3) \geq s_r + 2$. Abbott and Hanson [1972] showed that

$$s_{p+q} \geq 2s_p s_q + s_p + s_q.$$

Thus we have for fixed q and $r \geq q$:

$$R_r(3) \geq c(2s_q + 1)^{\frac{r}{q}},$$

where $c = c(q)$ is a constant depending only on q. Using that $s_5 \geq 157$ (Fredricksen (1979)) we get from this

$$R_r(3) \geq c(3.16)^r.$$

A lot of effort went into improving these bounds. The best known lower and upper bound, however, are still of the order c^r resp. $cr!$, i.e. of the same order as the bounds shown above. For references for these results as for results on other than triangle Ramsey numbers we refer to Chung and Grinstead (1983) and Radziszowski (2011).

Chapter 8
Rapidly Growing Ramsey Functions

Gödel's paper on formally undecidable propositions in first order Peano arithmetic
(Gödel 1931) showed that any recursive axiomatic system containing Peano arithmetic still admits propositions which are not decidable. Gödel's original example of
such a proposition was not that illuminating. It was merely a kind of formalization
of the well known antinomy of the liar. This raised the problem to look for
intuitively meaningful propositions which are independent of Peano arithmetic.
Paris and Harrington (1977) showed that a straightforward variant of the finite
Ramsey theorem is independent of Peano arithmetic, thus witnessing Gödel's first
incompleteness theorem.

The original short and elegant proof of Paris and Harrington uses model theoretic
tools. A different, purely combinatorial explanation of the unprovability by means of
fast growing functions was given by Ketonen and Solovay (1981). In this section we
present a simplification of the Ketonen-Solovay argument due to Loebl and Nešetřil
(1991). We start with some background on fast growing hierarchies.

8.1 The Hardy Hierarchy

Let $\gamma_1 = \omega$ and $\gamma_{n+1} = \gamma_n^{\omega}$ for every $n < \omega$, i.e.,

$$\gamma_n = \left. \omega^{\omega^{\cdot^{\cdot^{\cdot^{\omega}}}}} \right\} \text{n-times} .$$

Moreover set

$$\epsilon_0 = \omega^{\omega^{\cdot^{\cdot^{\cdot}}}} = \lim_{n \to \infty} \gamma_n.$$

Then ϵ_0 is the least ordinal solution to the equation $\omega^{\lambda} = \lambda$. Throughout this section
we are only concerned with ordinals below ϵ_0.

H.J. Prömel, *Ramsey Theory for Discrete Structures*,
DOI 10.1007/978-3-319-01315-2_8,
© Springer International Publishing Switzerland 2013

First note that every ordinal below ϵ_0 admits a unique representation known as the *Cantor normal form* of α:
Let $\alpha < \epsilon_0$ be a positive ordinal and k be a positive integer. Then α can be represented uniquely as

$$\alpha = \omega^{\alpha_1} \cdot n_1 + \omega^{\alpha_2} \cdot n_2 + \ldots + \omega^{\alpha_k} \cdot n_k,$$

where $\alpha > \alpha_1 > \alpha_2 > \ldots > \alpha_k \geq 0$ are ordinals and $n_1, \ldots, n_k$ are positive integers.

Such a coding of ordinals $\alpha < \epsilon_0$ by positive integers can be defined straightforwardly, compare for example Schütte (1977).

Next we define *fundamental sequences* which we will subsequently use in order to define the *Hardy hierarchy*. We need these fundamental sequences in order to handle limit ordinals properly. To every limit ordinal $\alpha < \epsilon_0$ we associate a strictly monotone sequence $\alpha[n]$, $n < \omega$, which approaches α from below. If $\alpha < \epsilon_0$ is given in Cantor normal form $\alpha = \alpha' + \omega^{\alpha_k} \cdot n_k$, where α_k is the minimal exponent, let

$$\alpha[n] = \begin{cases} \alpha' + \omega^{\alpha_k} \cdot (n_k - 1) + \omega^{\alpha_k[n]}, & \text{if } \alpha_k \text{ is a limit ordinal,} \\ \alpha' + \omega^{\alpha_k} \cdot (n_k - 1) + \omega^{\alpha_k - 1} \cdot (n + 1), & \text{if } \alpha_k \text{ is a successor ordinal.} \end{cases}$$

For example, $\omega[n] = n + 1$, $\omega^\omega[n] = \omega^{n+1}$, $\omega^{k+1}[n] = \omega^k \cdot (n + 1)$, and $\omega^k \cdot (k + 1)[n] = \omega^k \cdot k + \omega^{k-1} \cdot (n + 1)$.

With the help of these fundamental sequences we define functions $H_\alpha(\cdot)$ for all $\alpha < \epsilon_0$:

$$\begin{aligned} H_0(n) &= n, \\ H_{\alpha+1}(n) &= H_\alpha(n + 1), \\ H_\alpha(n) &= H_{\alpha[n]}(n) \qquad \text{for limit ordinals.} \end{aligned}$$

Finally, define H_{ϵ_0} by
$$H_{\epsilon_0}(n) = H_{\gamma_n}(n).$$

This is the *Hardy hierarchy*, introduced by Wainer (1972). This hierarchy is based on a sequence of functions first defined by Hardy (1904) to construct sets of real numbers of cardinality $\aleph_1$. It is not difficult to see that each H_α is strictly increasing and $H_\alpha(n) < H_{\alpha+1}(n)$ for every nonnegative integer n.

The significance of the Hardy hierarchy in connection with unprovability results stems from the following theorem, cf. Wainer (1970, 1972) and Buchholz and Wainer (1987).

Theorem 8.1. *Let $f : \omega \to \omega$ be a provably total and recursive function (provably total with respect to Peano arithmetic). Then f is eventually dominated by some H_α for an $\alpha < \epsilon_0$. Moreover, H_{ϵ_0} eventually dominates every provably total recursive function but it itself is not provably total.*

8.2 Paris-Harrington's Unprovability Result

A set $L \subseteq \omega$ is called large, if $L \neq \emptyset$ and $\min L \leq |L|$. So $\{4, 5, 6, 7\}$ is a large set but not $\{4, 10, 15\}$. Let k, n and r be positive integers. With this terminology at hand we can state the following variation of the classical Ramsey theorem that follows from the infinite Ramsey theorem using a compactness argument.

Theorem 8.2. *Let k and r be positive integers. Then there exists a least positive integer $n = PH(k, r)$ such that for every r-coloring $\Delta : [n]^k \rightarrow r$ there exists a large subset $L \subseteq n$ with $|L| > k$ such that $\Delta \rceil [L]^k$ is a constant coloring.* $\square$

While for the classical Ramsey theorem it is difficult to obtain *tight* bounds it will turn out that for this seemingly small variation of the classical Ramsey theorem it is already difficult to obtain *any* kind of bound.

Theorem 8.3 (Paris and Harrington). *The statement*

(PH) for every pair k, r of positive integers there exists a least positive integer $n = PH(k, r)$ such that for every r-coloring $\Delta : [n]^k \rightarrow r$ there exists a large subset $L \subseteq n$ with $|L| > k$ such that $\Delta \rceil [L]^k$ is a constant coloring

is not provable in Peano arithmetic.

For the reader who is not used to work in Peano arithmetic we mention that for statements about natural numbers Peano arithmetic is equivalent to the result of replacing the axiom of infinity by its negation in the usual axioms of Zermelo-Fraenkel set theory (see, e.g., Jech (1978) for these axioms). Obviously, the principle (PH) can be formulated in this theory. In this way Theorem 8.3 should be understood as: the formula of Peano arithmetic corresponding to the principle (PH) is not provable in Peano arithmetic.

Intuitively, a reason for the unprovability of (PH) in Peano arithmetic is that the function $PH(k, r)$ grows too rapidly. Recall that a recursive function $f : \omega \rightarrow \omega$ is provably recursive if one can show in Peano arithmetic that f is total, i.e., defined for all natural numbers. Now it turns out that the function $PH(k, k)$ grows faster than any provably recursive function f, i.e., $f(k) < PH(k, k)$ for all but finitely many k. However, by Theorem 8.2 the function $PH(k, k)$ is total, hence, (PH) is not provable in Peano arithmetic.

The aim of this section is to prove the Paris-Harrington result by purely combinatorial means following an approach of Ketonen and Solovay (1981). Here we follow a simplified approach by Loebl and Nešetřil (1991).

Let $\alpha < \epsilon_0$ be an ordinal and let $\alpha = \omega^{\alpha_1} \cdot n_1 + \omega^{\alpha_2} \cdot n_2 + \ldots + \omega^{\alpha_k} \cdot n_k$ be the Cantor normal form of α. Let $S_i(\alpha) = \omega^{\alpha_i} \cdot n_i$ be the ith summand in the Cantor normal form of α, let $C_i(\alpha) = n_i$ be the coefficient of the ith summand and let $E_i(\alpha) = \alpha_i$ be the corresponding exponent. If $\gamma_{h-1} \leq \alpha < \gamma_h$ then α is said to be of height h which is abbreviated by $h(\alpha) = h$. The weight $w(\alpha)$ of α is defined recursively as follows:

$$w(\alpha) = \begin{cases} \alpha, & \text{if } \alpha \text{ is an integer,} \\ \max\{n_1, \ldots, n_k, w(\alpha_1), \ldots, w(\alpha_k), k\}, & \text{otherwise.} \end{cases}$$

Let n be an integer. Then (α, n) is called a *good pair* if $n > w(\alpha) + h(\alpha)$. Let (α, n) be a good pair. We define a predecessor function $R(\alpha; n)$ as follows.

$$R(\alpha; n) = \begin{cases} (\alpha - 1; n + 1), & \text{if } \alpha \text{ is a successor ordinal,} \\ (\alpha[n - h(\alpha)]; n + 1), & \text{if } \alpha \text{ is a limit ordinal.} \end{cases}$$

Since $w(\alpha[n - h(\alpha)]) \leq \max\{w(\alpha) + 1, n - h(\alpha)\}$, it follows that $R(\alpha; n)$ again is a good pair. As an example, consider γ_h, the stack of h many ω's. Observe that $h(\gamma_h) = h + 1$. Hence, $(\gamma_h;\ h + 3)$ is a good pair and so is $R(\gamma_h;\ h + 3) = (\gamma_h[2];\ h + 4)$.

Let $R^0(\alpha; n) = (\alpha; n)$ and $R^{k+1}(\alpha; n) = R(R^k(\alpha; n))$. By $\mathcal{R}(\alpha; n)$ we denote the family of all pairs which can be generated by successively applying this predecessor operation, i.e., $\mathcal{R}(\alpha; n) = \{R^i(\alpha; n) \mid i < \omega\}$. Finally, let $r(\alpha, n) = |\mathcal{R}(\alpha; n)|$. This function can be related to the Hardy hierarchy.

Lemma 8.4. *Let $\alpha < \epsilon_0$ be an ordinal and let n be a non-negative integer. Then*

$$r(\alpha, n + h(\alpha)) \geq H_\alpha(n) - n.$$

Proof. We apply transfinite induction on α. Obviously, for every natural number k, $r(k, n)$ is the length of the sequence $(k, n), (k - 1, n + 1), \ldots, (0, n + k)$. Thus $r(k, n + 1) = k + 1 > H_k(n) - n = k$.

In the induction step we have either $\alpha + 1$ being a successor ordinal. i.e.,

$$r(\alpha + 1, n + h(\alpha)) = 1 + r(\alpha, n + h(\alpha) + 1)$$
$$\geq 1 + H_\alpha(n + 1) - n - 1$$
$$= H_{\alpha+1}(n) - n,$$

or α being a limit ordinal and therefore

$$r(\alpha,\ n + h(\alpha)) = r(\alpha[n],\ n + h(\alpha) + 1) \geq H_{\alpha[n]}(n) - n = H_\alpha(n) - n,$$

as claimed.   □

A family of good pairs is called a *good family*. If there is a member of such a family of height h and, moreover, the height of each member is at most h then this family is said to be a *good family of height h*. A good family $(\beta_0; n_0), \ldots, (\beta_{t-1}; n_{t-1})$ is *monotone* if $\beta_i > \beta_j$ and $n_i < n_j$ for every pair $0 \leq i < j < t$. For instance, $\mathcal{R}(\gamma_h; h + 3)$ is a monotone family of height $h + 1$ for every $h < \omega$.

The following coloring lemma plays the key rôle in the proof of the Paris-Harrington result.

Lemma 8.5. *Let $h \geq 2$ be an integer and let $S = \{(\beta_0; n_0), \ldots, (\beta_{t-1}; n_{t-1})\}$ be a good family of height h such that $n_i > h + 1$ for every $i < t$. Then there exists a coloring of the $(h + 1)$-subsets of S with less than 3^h colors such that no monotone subfamily $S' = \{(\alpha_0; m_0), \ldots, (\alpha_{s-1}; m_{s-1})\}$ of S of size $|S'| > m_0$ is monochromatic.*

Proof. Let $\epsilon_0 > \alpha_0 > \alpha_1 > \alpha_2$ be ordinals in Cantor normal form. Then let $\Delta(\alpha_0, \alpha_1) = \min\{i \mid S_i(\alpha_0) \neq S_i(\alpha_1)\}$ be the index of the largest summand where α_0 and α_1 differ. Recall that $C_i(\alpha_1), C_i(\alpha_2)$ denotes the coefficient of the ith summand of α_1, α_2, respectively. We define $\delta(\alpha_0, \alpha_1, \alpha_2) < 3$ as follows.

$$\delta(\alpha_0, \alpha_1, \alpha_2) = \begin{cases} 0, & \text{if } \Delta(\alpha_0, \alpha_1) > \Delta(\alpha_1, \alpha_2), \\ 1, & \text{if } x = \Delta(\alpha_0, \alpha_1) \leq \Delta(\alpha_1, \alpha_2) = y \text{ and } C_y(\alpha_1) < C_x(\alpha_0), \\ 2, & \text{otherwise.} \end{cases}$$

Iterating this scheme we associate to every strictly monotone decreasing sequence $\alpha = (\alpha_0, \ldots, \alpha_{t-1})$ of ordinals a vector $\delta(\alpha) = (\delta_0, \ldots, \delta_{t-3}) \in 3^{t-2}$ where $\delta_i = \delta(\alpha_i, \alpha_{i+1}, \alpha_{i+2})$.

Let $S = \{(\beta_0; n_0), \ldots, (\beta_{t-1}; n_{t-1})\}$ be a good family of height h such that $\beta_0 > \ldots > \beta_{t-1}$. We define a coloring of the $(h + 1)$-subsets of S by induction on h.

First assume that S is of height 2. Then color every monotone 3-element subset $\{(\alpha_0; m_0), (\alpha_1; m_1), (\alpha_2; m_2)\}$ of S with color $\delta(\alpha_0, \alpha_1, \alpha_2)$. This is clearly a 3-coloring. Assume that $S' = \{(\alpha_0; m_0), \ldots, (\alpha_{s-1}; m_{s-1})\}$ is a monotone subfamily of S which is monochromatic. Recalling the definition of $w(\alpha_0)$ and the fact that $m_0 > w(\alpha_0) + h(\alpha_0)$ we show in the following that $|S'| \leq m_0$. The assumption that S' is monochromatic with color 0 implies that $|S'|$ is bounded by the number of summands in the Cantor normal form of α_0 plus one. The assumption that S' is monochromatic with color 1 implies that $|S'|$ is at most one more than the size of the coefficient of the largest summand in the Cantor normal form of α_0. Finally, the assumption that S' is monochromatic with color 2 implies that $|S'|$ is bounded by the size of the exponent of the first summand in the Cantor normal form of α_0 plus one. This is because $h(\alpha_0) = 2$, i.e., the exponent is an integer.

Next assume the validity of the lemma for all good families of height h for some $h \geq 2$ and assume that S is of height $h + 1$ and therefore $n_i > h + 2$ for every $i < t$.

We associate a family $H(S)$ of height h to S as follows. To any 2-subset of S, say $\{(\alpha_0; m_0), (\alpha_1; m_1)\}$, we associate a pair $(\eta_0; p_0)$ choosing $p_0 = m_0 - 1$ and $\eta_0 = E_x(\alpha_0)$ where $x = \Delta(\alpha_0, \alpha_1)$. Observe that each such pair is a good pair of height at most h. Let $H(S)$ be the set of all pairs which can be obtained this way. Then $H(S)$ is a good family and $p > h + 1$ for every pair $(\eta, p) \in H(S)$ is valid. Without loss of generality we can assume that $H(S)$ is of height h. Hence by inductive assumption there exists a coloring of the $(h+1)$-subsets of $H(S)$ with less than 3^h colors such that no monotone subfamily $H' = \{(\eta_0; p_0), \ldots, (\eta_{r-1}; p_{r-1})\}$ of $H(S)$ of size $|H'| > p_0$ is monochromatic.

Now color the monotone $(h + 2)$-subfamilies of S as follows. Let $T = \{(\alpha_0; m_0), \ldots, (\alpha_{h+1}; m_{h+1})\}$ be such a family. Color T with $\delta(\alpha_0, \ldots, \alpha_{h+1}) \in 3^h$ if $\delta(\alpha_0, \ldots, \alpha_{h+1}) \neq (2, \ldots, 2)$. Otherwise consider the $(h+1)$-subfamily $H(T) = \{(\eta_0; p_0), \ldots, (\eta_h; p_h)\}$ of $H(S)$ and color T with the color assigned to $H(T)$ by the inductive assumption.

Obviously, this defines a coloring of all $(h + 2)$-subfamilies of S with less than $2 \cdot 3^h < 3^{h+1}$ many colors. Assume that $S' = \{(\alpha_0; m_0), \ldots, (\alpha_{s-1}; m_{s-1})\}$ is a monotone subfamily of S which is monochromatic. Assume that S' is monochromatic in some color $\delta \in 3^h$ which is not a constant vector. Then $|S'| \leq h + 2$, but $m_0 > h + 2$. If S' is monochromatic with color $(0, \ldots, 0) \in 3^h$ or with color $(1, \ldots, 1) \in 3^h$ similar arguments as in the case $h = 2$ show that $|S'| \leq w(\alpha_0) + 1$ but $m_0 > w(\alpha_0) + h(\alpha_0)$. It remains to consider the case that S' is monochromatic with color $(2, \ldots, 2) \in 3^h$. But then the family $H(S') = \{(\eta_0; p_0), \ldots, (\eta_{s-2}; p_{s-2})\}$, where $p_i = m_i - 1$ and $\eta_i = E_x(\alpha_i)$ with $x = \Delta(\alpha_0, \alpha_1)$ for every $i \leq s - 2$, is a monotone subfamily of $H(S)$ and, by definition of the coloring of S, monochromatic. Hence, by inductive assumption, $|H(S')| \leq p_0 = m_0 - 1$ and so $|S'| \leq m_0$. $\qquad\square$

Lemma 8.6. *Let $h \geq 2$. Then*

$$PH(h + 2, \, 3^{h+1} + 2h) \; > \; H_{\gamma_h}(h) + h.$$

Proof. Consider the monotone family $\mathcal{R}(\gamma_h; 2h + 1) = \{(\alpha_0; m_0), \ldots, (\alpha_{t-1}; m_{t-1})\}$. Obviously, $\alpha_0 = \gamma_h$, $m_0 = 2h + 1$ and $m_{i+1} = m_i + 1$ for every $i < t - 1$. By Lemma 8.4 we have that $t \geq H_{\gamma_h}(h) - h$. Since $\mathcal{R}(\gamma_h; 2h + 1)$ is a monotone family of height $h + 1$ and $m_i > h + 2$ for every $i < t$, by Lemma 8.5 there exists a coloring of the $(h + 2)$-subsets of $\mathcal{R}(\gamma_h; 2h + 1)$ with less than 3^{h+1} colors such that no monotone subfamily $S' \subseteq S$ of size $|S'| > 2h + 1$ is monochromatic. This induces obviously a coloring of the $(h + 2)$-subsets of $M = \{0, \ldots, 2h, 2h+1, \ldots, m_{t-1}\}$ with $3^{h+1}+2h$ many colors having the property that there is no large subset of L which is monochromatic. Hence

$$PH(h + 2, \, 3^{h+1} + 2h) \geq m_{t-1} + 1 \geq 2h + t + 1$$

$$\geq H_{\gamma_h}(h) + h + 1.$$

$\qquad\square$

Proof of Theorem 8.3.

$$PH(h + 2, \, 3^{h+1} + 2h) \; > \; H_{\gamma_h}(h) + h + 1 \; \geq \; H_{\epsilon_0}(h),$$

and so, by Theorem 8.1, $PH(h, h)$ is not a provably total and recursive function.

$\qquad\square$

It should be mentioned that other variants of Ramsey-type theorems give rise to functions which grow even much faster than the Paris-Harrington function. For example, in Prömel et al. (1991) fast growing functions based on Ramsey's theorem are investigated which grow faster than any recursive function which can proved to be total in the formal system ATR_0.

Chapter 9
Product Theorems

In this section we investigate product Ramsey theorems. Recall that the pigeonhole principle implies that if we color $r(m-1)+1$ points with r many colors, then at least one color class contains m points. Ramsey's theorem generalizes this from points to k-subsets. Another generalization of the pigeonhole principle is from points to pairs of points:

Proposition 9.1. *Let m and r be positive integers. Then there exits an integer n such that for every r-coloring $\Delta : n \times n \rightarrow r$ there exist subsets $A \in [n]^m$ and $B \in [n]^m$ such that $\Delta \restriction A \times B$ is a constant coloring.*

Proof. Let $n_0 = r(m-1)+1$ and $n_1 = r \cdot \binom{n_0}{m} \cdot (m-1)+1$. Consider a coloring $\Delta : n_0 \times n_1 \rightarrow r$. By the pigeonhole principle, for every $i < n_1$ there exists a set $A_i \in [n_0]^m$ such that $\Delta \restriction A_i \times \{i\}$ is a constant coloring. Applying the pigeonhole principle once again on the coloring $\Delta' : n_1 \rightarrow \binom{n_0}{m} \cdot r$, given by $\Delta'(i) = \langle A_i, \Delta(A_i \times \{i\})\rangle$, there exists an m-element set $B \in [n_1]^m$ such that Δ' is constant on B. This in particular implies that for all $i, j \in B$ we have $A_i = A_j$ and all the restrictions $\Delta \restriction A_i \times \{i\}$ are constant in the same color. Choosing $n = n_1$ thus completes the proof. □

Erdős and Rado (1956) invented the so-called polarized partition arrow to abbreviate such product situations. The special case of Proposition 9.1, for example, is abbreviated by

$$\binom{n}{n} \rightarrow \binom{m}{m}_r^{1,1}.$$

In Sect. 9.1 we prove a finite product Ramsey theorem of the following form. Let m, r, t and $k_0, \ldots, k_{t-1}$ be positive integers. Then there exist positive integers $n_0, \ldots, n_{t-1}$ such that

H.J. Prömel, *Ramsey Theory for Discrete Structures*,
DOI 10.1007/978-3-319-01315-2_9,
© Springer International Publishing Switzerland 2013

$$
\begin{pmatrix} n_0 \\ n_1 \\ \vdots \\ n_{t-1} \end{pmatrix} \rightarrow \begin{pmatrix} m \\ m \\ \vdots \\ m \end{pmatrix}_r^{k_0,k_1,\dots,k_{t-1}},
$$

meaning that for every r-coloring $\Delta : [n_0]^{k_0} \times [n_1]^{k_1} \times \dots \times [n_{t-1}]^{k_{t-1}} \rightarrow r$ there exists sets $A_i \in [n_i]^m$, for $i < t$, such that $\Delta][A_0]^{k_0} \times [A_1]^{k_1} \times \dots \times [A_{t-1}]^{k_{t-1}}$ is monochromatic.

In Sect. 9.2 we introduce the concept of diversification dealing with several unrestricted colorings acting on the same set. This concept turned out to be quite useful. As an application we deduce in Sect. 9.3 a product version of the finite Erdős-Rado canonization theorem originally due to Rado (1954).

9.1 A Product Ramsey Theorem

In the terminology of graph theory a rectangle $A \times B \in [n_0]^{k_0} \times [n_1]^{k_1}$ corresponds to a K_{k_0,k_1}-subgraph of the complete bipartite graph K_{n_0,n_1}. The product Ramsey theory in this special case $t = 2$ thus corresponds to the question: suppose we color K_{k_0,k_1}-subgraphs of the complete bipartite graph K_{n_0,n_1}, can we find a monochromatic $K_{m,m}$-subgraph. The following theorem shows that this is indeed true, whenever n is large enough.

Theorem 9.2 (Product Ramsey theorem). *Let t, $(k_i)_{i<t}$, m and r be positive integers. Then there exists a positive integer $n = n((k_i)_{i<t}, m, r)$ such that for every coloring $\Delta : \prod_{i<t}[n]^{k_i} \rightarrow r$ there exist m-subsets $(M_0, \dots, M_{t-1}) \in \prod_{i<t}[n]^m$ such that*

$$
\Delta(A_0, \dots, A_{t-1}) = \Delta(B_0, \dots, B_{t-1}),
$$

for all $(A_0, \dots, A_{t-1}), (B_0, \dots, B_{t-1}) \in \prod_{i<t}[M_i]^{k_i}$.

Proof. We proceed by induction on t, the case $t = 1$ being Ramsey's theorem.

Let n be according to the inductive hypothesis with respect to $(k_i)_{i<t}, m$ and r, i.e.,

$$
\begin{pmatrix} n \\ n \\ \vdots \\ n \end{pmatrix} \rightarrow \begin{pmatrix} m \\ m \\ \vdots \\ m \end{pmatrix}_r^{k_0,k_1,\dots,k_{t-1}},
$$

and choose N according to Ramsey's theorem such that $N \rightarrow (m)^{k_t}_{r^p}$, where $p = \prod_{i<t}[n]^{k_i}$.

Now let $\Delta : \prod_{i<t}[n]^{k_i} \times [N]^{k_t} \rightarrow r$ be a coloring. We define $\Delta_t : [N]^{k_t} \rightarrow r^p$ by

$$\Delta_t(K_t) = \langle \Delta(K_0, \ldots, K_{t-1}, K_t) \mid (K_0, \ldots, K_{t-1}) \in \prod_{i<t}[n]^{k_i} \rangle.$$

By choice of N there exists $M_t \in [N]^m$ such that $\Delta_t \upharpoonright [M]^{k_t}$ is constant, which is to say that $\Delta \upharpoonright (\prod_{i<t}[n]^{k_i} \times [M_t]^{k_t})$ is independent of the tth coordinate. Hence, by inductive hypothesis we get an $(M_0, \ldots, M_t) \in \prod_{i<t}[n]^m \times [N]^m$ monochromatic with respect to Δ.   □

Notice that Theorem 9.2 remains valid if (at most) in one coordinate the m (and thus the n) is replaced by ω. However, even for $k = 1$ it becomes false if (at least) in two of the coordinates the m are replaced by ω, as the following example shows.

Let $\Delta : \omega \times \omega \rightarrow 2$ be given by

$$\Delta(x, y) = \begin{cases} 0, & \text{if } x \leq y \\ 1, & \text{otherwise.} \end{cases}$$

Then, obviously, no pair $(F_0, F_1) \in [\omega]^\omega \times [\omega]^\omega$ is colored monochromatically.

9.2 Diversification

Let $k \leq \ell$ and $\Delta_0 : [n]^k \rightarrow \omega$, $\Delta_1 : [n]^\ell \rightarrow \omega$ be colorings for some n sufficiently large. Then according to the Erdős-Rado canonization theorem (applied twice) there exists $M \in [n]^m$ such that $\Delta_0 \upharpoonright [M]^k$ as well as $\Delta_1 \upharpoonright [M]^\ell$ are canonical colorings. But in this way we do not get any information about dependencies between the colors used by $\Delta_0 \upharpoonright [M]^k$ and $\Delta_1 \upharpoonright [M]^\ell$. To obtain such information we introduce the concept of diversification:

Theorem 9.3. *Let $k \leq \ell$ and m be positive integers. Then there exists a positive integer n such that for each pair $\Delta_0 : [n]^k \rightarrow \omega$ and $\Delta_1 : [n]^\ell \rightarrow \omega$ of colorings there exists an $M \in [n]^m$ and there exists a pair $J_0 \subseteq k$ and $J_1 \subseteq \ell$ of sets such that*

(1) $\Delta_0 \upharpoonright [M]^k$ *is canonical with respect to* J_0,
     $\Delta_1 \upharpoonright [M]^\ell$ *is canonical with respect to* J_1, *and*
(2) *Either* $\Delta_0(A) \neq \Delta_1(B)$ *for all* $A \in [M]^k$ *and* $B \in [M]^\ell$
     or $\Delta_0(A) = \Delta_1(B)$ *if and only if* $A : J_0 = B : J_1$ *for all* $A \in [M]^k$ *and* $B \in [M]^\ell$.

Diversification, i.e., separating different colorings, was developed in Voigt (1985). In fact, more general results than Theorem 9.3 are true in this direction.

The key in proving the theorem is the following lemma for one-to-one colorings.

Lemma 9.4. *Let* $i \leq j$ *and* m *be positive integers. Then there exists a positive integer* n *such that for each pair* $\Delta_0 : [n]^i \rightarrow \omega$ *and* $\Delta_1 : [n]^j \rightarrow \omega$ *of one-to-one colorings there exists* $M \in [n]^m$ *such that one of the following possibilities holds:*

(1) $\Delta_0(A) \neq \Delta_1(B)$ *for all* $A \in [M]^i$, $B \in [M]^j$,
(2) $i = j$ *and* $\Delta_0(A) = \Delta_1(A)$ *for all* $A \in [M]^i$.

Proof. Let $m' = m + j - i$ and choose m^* such that $m^* \rightarrow (m')^j_3$. Finally, choose n such that $n \rightarrow (m^*)^j_2$.

Now assume Δ_0, Δ_1 are given as stated in the lemma. Recall that $i \leq j$ and that Δ_0 is defined on i-subsets of n. We extend Δ_0 to j-subsets of n as follows. Let $\Delta^1_0 : [n]^j \rightarrow \omega$ be defined by $\Delta^1_0(X) = \Delta_0(\{x_0, \ldots, x_{i-1}\})$, where $x_0, \ldots, x_{i-1}$ are the first i elements of X with respect to the natural order of n. Now define a coloring $\Delta^* : [n]^j \rightarrow 2$ by

$$\Delta^*(X) = \begin{cases} 1, & \text{if } \Delta^1_0(X) = \Delta_1(X) \\ 0, & \text{otherwise.} \end{cases}$$

By choice of n there exists $M^* \in [n]^{m^*}$ such that $\Delta^* \lceil [M^*]^j$ is a constant coloring. In case $\Delta^* \lceil [M^*]^j \equiv 1$ it follows from the fact that Δ_1 is one-to-one that necessarily $i = j$ and M^* thus satisfies (2).

So assume that $\Delta^* \lceil [M^*]^j \equiv 0$. Then we impose a directed graph on $[M^*]^j$ letting (X, Y) be an edge if $\Delta^1_0(X) = \Delta_1(Y)$. Clearly this graph has no loops and, since Δ_1 is one-to-one, the outdegree of every vertex is at most one. Therefore each connected component of this graph contains at most one cycle and hence, the underlying undirected graph is 3-colorable.

Given such a 3-coloring, by choice of m^* there exists a monochromatic m'-set $M' \in [M^*]^{m'}$. Choosing M as the first m elements of M' satisfies (1). $\square$

Proof of Theorem 9.3. Let n' be such that the above lemma can be applied for every pair $i \leq k$ and $j \leq \ell$ and m. Further, let n be such that after applying the Erdős-Rado canonization theorem to colorings $\Delta_0 : [n]^k \rightarrow \omega$ and $\Delta_1 : [n]^\ell \rightarrow \omega$, we may assume that $\Delta_0 \lceil [n']^k$ and $\Delta_1 \lceil [n']^\ell$ are canonical colorings with respect to some $J_0 \subseteq k$ and $J_1 \subseteq \ell$, respectively.

Let $\Delta^*_0 : [n']^{|J_0|} \rightarrow \omega$, resp. $\Delta^*_1 : [n']^{|J_1|} \rightarrow \omega$, be such that

$$\Delta^*_0(A : J_0) = \Delta_0(A) \qquad \text{and} \qquad \Delta^*_1(B : J_1) = \Delta_1(B).$$

Observe that the assumption that $\Delta_0(A) = \Delta_0(B)$ if and only if $A : J_0 = B : J_0$ implies that Δ^*_0 is well defined and one-to-one. (For sets $A^* \in [n']^{|J_0|}$ that cannot be written in the form $A : J_0$ we define $\Delta^*_0(A^*)$ arbitrarily, but so that the function remains one-to-one.) Similarly, we deduce that Δ^*_1 is well-defined and one-to-one. Applying Lemma 9.4 we find some $M \in [n']^m$. If M satisfies property (1) of

Lemma 9.4, then we have for any $A \in [M]^k$ and $B \in [M]^\ell$ that $\Delta(A) = \Delta_0^*(A : J_0) \neq \Delta_1^*(B : J_1) = \Delta(B)$. Otherwise, from property (2) and the fact that Δ_0^* and Δ_1^* are one-to-one we have $\Delta(A) = \Delta_0^*(A : J_0) = \Delta_1^*(B : J_1) = \Delta(B)$ if and only if $A : J_0 = B : J_1$. Therefore M satisfies the theorem. $\square$

Lemma 9.4 was independently obtained by Meyer auf der Heide and Wigderson (1987) in proving lower bounds for sorting networks. We have adopted some of their ideas here.

9.3   A Product Erdős-Rado Theorem

A t-dimensional version of the Erdős-Rado canonization theorem was established in Rado (1954). Loosely speaking it asserts that in each coordinate we have a canonical coloring.

Theorem 9.5. *Let t, $(k_i)_{i<t}$ and m be positive integers. Then there exists a positive integer $n = n((k_i)_{i<t}, m)$ such that for every coloring $\Delta : \prod_{i<t}[n]^{k_i} \to \omega$ there exist m-subsets $(M_0, \ldots, M_{t-1}) \in \prod_{i<t}[n]^m$ and there exist (possibly empty) sets $J_i \subseteq k_i$ for $i < t$ such that*

$$\Delta(A_0, \ldots, A_{t-1}) = \Delta(B_0, \ldots, B_{t-1})$$

if and only if $A_i : J_i = B_i : J_i$ for every $i < t$,

for all $(A_0, \ldots, A_{t-1})$ and $(B_0, \ldots, B_{t-1}) \in \prod_{i<t}[M_i]^{k_i}$.

Proof. We proceed by induction on t, the case $t = 1$ being the Erdős-Rado canonization theorem. Let m^* be according to the inductive hypothesis with respect to $(k_i)_{i<t}$ and m. Furthermore, choose n according to the product Ramsey theorem such that

$$\begin{pmatrix} n \\ n \\ \vdots \\ n \end{pmatrix} \to \begin{pmatrix} m^* \\ m^* \\ \vdots \\ m^* \end{pmatrix}_{2^{k_t}}^{k_0, k_1, \ldots, k_{t-1}} .$$

Finally, choose N large enough so that Theorem 9.3 can be applied successively $\binom{\prod_{i<t}\binom{n}{k_i}}{2}$-times for colorings acting on k_t-sets, and yielding a set of size m after the last application of Theorem 9.3.

Let $\Delta : \prod_{i<t}[n]^{k_i} \times [N]^{k_t} \to \omega$ be a coloring. For every $\mathcal{K} = (K_0, \ldots, K_{t-1})$, where $K_i \in [n]^{k_i}$ for $i < t$, let $\Delta_{\mathcal{K}} : [N]^k \to \omega$ be given by $\Delta_{\mathcal{K}}(K_t) = \Delta(K_0, \ldots, K_{t-1}, K_t)$. By choice of N there exists $M_t \in [N]^m$ such that for every pair $\Delta_{\mathcal{K}}, \Delta_{\mathcal{K}'}$ the assertion of Theorem 9.3 is valid.

Observe that property (1) of Theorem 9.3 implies that for every $\mathcal{K}$ there exits a set $J_{\mathcal{K}} \subseteq k_t$ such that $\Delta_{\mathcal{K}} \rceil [M_t]^{k_t}$ is canonical with respect to $J_{\mathcal{K}}$. Define a coloring $\Delta^* : \prod_{i<t} [n]^{k_i} \to 2^{k_t}$ such that $\Delta^*(\mathcal{K}) = J_{\mathcal{K}}$ for every $\mathcal{K} = (K_0, \ldots, K_{t-1})$. By choice of n we can apply the product Ramsey theorem to find $(M_0^*, \ldots, M_{t-1}^*) \in \prod_{i<t} [n]^{m^*}$ such that there exists just one $J_t \subseteq k_t$ so that for every $\mathcal{K} \in \prod_{i<t} [M_i^*]^{k_i}$, it follows that $\Delta_{\mathcal{K}}(A) = \Delta_{\mathcal{K}}(B)$ if and only if $A : J_t = B : J_t$, whenever $A, B \in [M_t]^{k_t}$.

Finally, define a coloring $\Delta^{**} : \prod_{i<t} [M_i^*]^{k_i} \to \omega$ such that

$$\Delta^{**}(A_0, \ldots, A_{t-1}) = \Delta^{**}(B_0, \ldots, B_{t-1}) \quad \text{if and only if}$$

$$\Delta(A_0, \ldots, A_{t-1}, K_t) = \Delta(B_0, \ldots, B_{t-1}, K_t) \quad \text{for some } K_t \in [M_t]^{k_t}.$$

Observe that by property (2) of Theorem 9.3, Δ^{**} is well-defined. Then by induction hypothesis there exists $(M_0, \ldots, M_{t-1}) \in \prod_{i<t} [n]^m$ and there exist $J_i \subseteq k_i, i < t$, such that $\Delta^*(A_0, \ldots, A_{t-1}) = \Delta^*(B_0, \ldots, B_{t-1})$ if and only if $A_i : J_i = B_i : J_i$ for every $i < t$, for all $(A_0, \ldots, A_{t-1})$ and $(B_0, \ldots, B_{t-1})$ from $\prod_{i<t} [M_i]^{k_i}$.

An easy calculation shows that $(M_0, \ldots, M_{t-1}, M_t)$ and $J_0, \ldots, J_{t-1}, J_t$ satisfy Theorem 9.5. $\square$

For more general product theorems compare, e.g., Graham and Spencer (1979) and Voigt (1985). Here we just write down the special case of Theorem 9.5 when all $k_i = 1$. This is the t-dimensional canonical pigeonhole principle.

Corollary 9.6. *Let t and m be positive integers. Then there exists a least positive integer $n = n(m, t)$ such that for every coloring $\Delta : [n]^t \to \omega$ there exist subsets $M_i \in [n]^m$, $i < t$, and there exists a (possibly empty) set $J \subseteq t$ such that*

$$\Delta(a_0, \ldots, a_{t-1}) = \Delta(b_0, \ldots, b_{t-1}) \quad \text{if and only if} \quad a_j = b_j \text{ for all } j \in J$$

for all $(a_0, \ldots, a_{t-1}), (b_0, \ldots, b_{t-1}) \in \prod_{i<t} M_i$. $\square$

Chapter 10
A Quasi Ramsey Theorem

The basic problem of (combinatorial) discrepancy theory is how to color a set with two colors as uniformly as possible with respect to a given family of subsets. The aim is to achieve that each of the two colors meets each subset under consideration in approximately the same number of elements. From the finite Ramsey theorem (cf. Corollary 7.2) we know already that if the set of all 2-subsets of n is 2-colored, and the family of all ℓ-subsets for some $\ell < \frac{1}{2}\log n$ is considered, the situation is as bad as possible: for any 2-coloring we will find a monochromatic ℓ-set. As ℓ gets larger one can color more uniformly though one still has the preponderance phenomenon.

Let k and n be positive integers and let $\chi_k : [n]^k \to \{-1, +1\}$ be a 2-coloring of the k-subsets of n. For $T \subseteq n$ let

$$\chi_k(T) = \sum_{X \in [T]^k} \chi_k(X).$$

Then $\chi_k(T) = 0$ means that T is colored as uniformly as possible, i.e., the color '-1' and the color '$+1$' occur equally often. The discrepancy of χ_k is defined by

$$\operatorname{disc}(\chi_k) = \max_{T \subseteq n} |\chi_k(T)|.$$

and the discrepancy of n with respect to colorings of k-subsets is given by

$$\operatorname{disc}(k, n) = \min \operatorname{disc}(\chi_k),$$

where the minimum is taken over all 2-colorings $\chi_k : [n]^k \to \{-1, +1\}$. Trivially, $\operatorname{disc}(1, n) = \lceil \frac{n}{2} \rceil$ for every n. From Corollary 7.2 we also get that $\operatorname{disc}(2, n) > \frac{1}{2}\log n$.

H.J. Prömel, *Ramsey Theory for Discrete Structures*,
DOI 10.1007/978-3-319-01315-2_10,
© Springer International Publishing Switzerland 2013

Extending earlier results of Erdős (1963) and Erdős and Spencer (1972) proved:

Theorem 10.1 (Erdős, Spencer). *Let k be a positive integer. Then there exist constants $c_0 = c_0(k)$ and $c_1 = c_1(k)$ such that for every n*

$$c_0 n^{\frac{k+1}{2}} \leq \mathrm{disc}(k, n) \leq c_1 n^{\frac{k+1}{2}}$$

In this section we will focus on the discrepancy problem for finite sets, i.e., on Theorem 10.1. For an excellent surveys on discrepancy results in general see e.g. Sós (1983) and Beck and Sós (1995) or the book by Chazelle (2000).

10.1   The Upper Bound

It is not surprising that the upper bound in Theorem 10.1 is given by probabilistic means. The basic tool in proving this upper bound is the inequality of Chernoff (1952). Here we use it in a version given by Spencer (1985, p. 362).

Lemma 10.2 (Chernoff). *Let X_i, $i < n$, be mutually independent random variables with $\mathrm{Prob}[X_i = -1] = \mathrm{Prob}[X_i = +1] = \frac{1}{2}$ for $i < n$ and put $S_n = \sum_{i<n} X_i$. Let $a > 0$ be some constant. Then*

$$\mathrm{Prob}[S_n > a] < e^{-\frac{1}{2}\frac{a^2}{n}}.$$

$\square$

Now fix some $k \geq 1$ and let $\chi_k : [n]^k \to \{-1, +1\}$ be a random mapping, taking the values -1 and $+1$ each with probability $\frac{1}{2}$ and independently. For each $T \subseteq n$ the distribution of $\chi_k(T)$ is the same as that of $S_{\binom{|T|}{k}}$ and therefore, by Chernoff's lemma,

$$\mathrm{Prob}[|\chi_k(T)| > cn^{\frac{k+1}{2}}] < 2\exp\left(\frac{-c^2 n^{k+1}}{2\binom{|T|}{k}}\right) < 2\exp\left(\frac{-c^2 n^{k+1}}{2n^k}\right) = 2e^{-\frac{c^2}{2}n}.$$

Since there are 2^n choices for T we get

$$\mathrm{Prob}[\max_{T \subseteq n} |\chi_k(T)| > cn^{\frac{k+1}{2}}] \leq 2^{n+1} e^{-\frac{c^2}{2}n} < 1,$$

choosing, e.g., $c = c_1 < \sqrt{2\ln 2} + 1$. So there exists $\chi_k : [n]^k \to \{-1, +1\}$ such that $\max_{T \subseteq n} |\chi_k(T)| \leq c_1 n^{\frac{k+1}{2}}$ and, hence, $\mathrm{disc}(k, n) \leq c_1 n^{\frac{k+1}{2}}$. $\square$

10.2 A Lemma of Erdős

In connection with his investigations on a lemma of Littlewood and Offord (1943) and Erdős (1945) proved the following result.

Lemma 10.3. *Let $x_0, \ldots, x_{n-1}$ be reals satisfying $|x_i| \geq 1$ for every $i < n$. Then for every $r \in \mathbb{R}$ the number of sums $\sum_{i<n} \epsilon_i x_i$, where $\epsilon_i \in \{0, +1\}$, which fall into the (halfopen) interval $[r, r + 1[$ does not exceed $\binom{n}{\lfloor \frac{n}{2} \rfloor}$.*

Proof. We first show that it suffices to consider the case that the x_i are all nonnegative. Indeed, assume that $x_i < 0$ for some $i < n$. If we replace x_i by $-x_i$ and each ϵ_i by $(\epsilon_i + 1) \bmod 2$, then all sums are shifted by exactly $-x_i$. The lemma thus follows by considering the case $r - x_i$.

So assume that $x_i \geq 1$ for every i. Now for every sum $\sum_{i<n} \epsilon_i x_i$, the ϵ_i can be viewed as the characteristic function of a subset of n. If $\sum_{i<n} \epsilon x_i$ and $\sum_{i<n} \eta_i x_i$ are both in $[r, r + 1[$, for some $r \in \mathbb{R}$, then neither of the corresponding subsets contains the other. Hence, by Sperner's lemma (Sperner 1928), the number of sums which fall in the interval $[r, r + 1[$ does not exceed $\binom{n}{\lfloor \frac{n}{2} \rfloor}$. $\qquad\square$

What we actually need in order to prove Theorem 10.1 is the following corollary of Lemma 10.3:

Corollary 10.4. *There exists a positive integer n_0 such that for every $n \geq n_0$, for every $0 < c \leq 1$ and for every sequence $x_0, \ldots, x_{n-1}$ of reals satisfying $|x_i| \geq 1$ for at least cn many $i < n$ we have that*

$$\left| \sum_{j \in J} x_j \right| \geq c \frac{\sqrt{n}}{2}, \tag{10.1}$$

for at least $\frac{1}{5} 2^n$ choices of $J \subseteq n$.

Proof. Let $I \subseteq n$ be such that $|x_i| \geq 1$ for every $i \in I$ and such that $|I| \geq cn$. Let $J \subseteq n$. If (10.1) does not hold then

$$-\sum_{j \in J \setminus I} x_j - c \frac{\sqrt{n}}{2} < \sum_{j \in I \cap J} x_j < -\sum_{j \in J \setminus I} x_j + c \frac{\sqrt{n}}{2}.$$

Now we think this open interval to be covered with $\lceil c\sqrt{n} \rceil$ halfopen intervals of length 1. Then assuming $J \setminus I$ to be fixed for the moment, by Erdős' lemma the assertion (10.1) is not fulfilled for at most

$$\lceil c\sqrt{n} \rceil \binom{cn}{\lfloor \frac{cn}{2} \rfloor} < \frac{4}{5} 2^{cn}$$

choices of $I \cap J$. (The inequality follows from $\binom{x}{x/2} = (1 + o(1))\sqrt{2/(\pi x)}2^x$.) Summing over all possible $J \setminus I$ (at most $2^{(1-c)n}$ many) yields the corollary. $\square$

Note that Erdős (1945) proved already that for any sequence of reals $x_0, \ldots, x_{n-1}$ with $|x_i| \geq 1$ the number of sums $\sum_{i<n} \epsilon_i x_i$ which fall into the interior of any interval of length $2\,m$, for some positive integer m, is not greater than the sum of the m greatest binomial coefficients. This, of course, allows to strengthen Corollary 10.4 considerably, but this is not of use for our purposes.

10.3   The Lower Bound: The Graph Case

Because of its particular interest and since its proof becomes considerably easier, we separate the graph case, i.e., the case $k = 2$.

Proposition 10.5. *There exist constants c_0 and c_1 so that for every n*

$$c_0 n^{3/2} \leq \mathrm{disc}(2, n) \leq c_1 n^{3/2}.$$

Proof. The upper bound was proven in Sect. 10.1, so we concentrate on the lower bound. Interpreting the lower bound in terms of graphs, Proposition 10.5 says that for every graph $G = (n, E)$ there exists an (induced) subgraph which has considerably more edges, viz. $c_0 n^{3/2}$, than non-edges, or vice versa. Assume that every edge has weight $+1$ and every non-edge has weight -1, which defines some $\chi : [n]^2 \to \{-1, +1\}$. Let $A_0, A_1 \subseteq n$ be disjoint subsets of n. Then, by abuse of language, we put

$$\chi(A_0, A_1) = \sum \chi(e),$$

where the summation is taken over all edges having one endpoint in A_0 and the other endpoint in A_1. Now we prove the lower bound proceeding in two steps. First we show:

There exists $\epsilon > 0$ such that for every $n \geq 2n_0$ (without loss of generality n is even), for every $\chi : [n]^2 \to \{-1, +1\}$ and every pair $A_0, A_1 \subseteq n$ of disjoint sets satisfying $|A_0| = |A_1| = \frac{n}{2}$, there exist $B_0 \subseteq A_0$, and $B_1 \subseteq A_1$ so that

$$|\chi(B_0, B_1)| \geq \epsilon n^{3/2}.$$

In order to prove this fix some $a \in A_1$. By Corollary 10.4 for $c = 1$, we have that

$$\left| \{ B \subseteq A_0 \mid |\chi(B, a)| \geq \frac{\sqrt{n}}{2\sqrt{2}} \} \right| \geq \frac{1}{5} 2^{n/2}.$$

Thus putting $\delta = \frac{1}{20}$ we obtain the existence of $B_0 \subseteq A_0$ satisfying

$$|\{a \in A_1 \mid |\chi(B_0, a)| \geq \frac{\sqrt{n}}{2\sqrt{2}}\}| \geq 2\delta n.$$

By symmetry we can assume that

$$\left|\{a \in A_1 \mid \chi(B_0, a) \geq \frac{\sqrt{n}}{2\sqrt{2}}\}\right| \geq \delta n.$$

Now let $B_1 = \{a \in A_1 \mid |\chi(B_0, a)| \geq \frac{\sqrt{n}}{2\sqrt{2}}\}$. Then

$$\chi(B_0, B_1) = \sum_{a \in B_1} \chi(B_0, a) \geq \delta n \frac{\sqrt{n}}{2\sqrt{2}} = \epsilon n^{3/2},$$

choosing $\epsilon = \frac{\delta}{2\sqrt{2}}$, thus our claim.

In a second step we have to transfer the imbalance of the bipartite graph into an imbalance of some subgraph. For this purpose let $\hat{c}_0 = \frac{\epsilon}{3}$ and observe that

$$\chi(B_0, B_1) = \chi(B_0 \cup B_1) - \chi(B_0) - \chi(B_1).$$

Thus, by the pigeonhole principle, either B_0, or B_1, or $B_0 \cup B_1$, has a discrepancy of size at least $\hat{c}_0 n^{3/2}$. Choosing $c_0 \leq \hat{c}_0$ to take care of the n's smaller than n_0 completes the proof of Proposition 10.5.   □

10.4   The Lower Bound: The General Case

The general approach for the case $k \geq 2$ is similar as in the graph case. First we aim at finding k pairwise disjoint subsets $A_0, \ldots, A_k$ such that the collection of all k-subsets that meat each of the A_i exactly once have a high discrepancy. In a second step we then argue that this implies the existence of a set A' that has a high discrepancy. The main idea is similar to the graph case. Differences arise mainly from the fact that given pairwise disjoint sets $A_0, \ldots, A_{k-1}$ there are many more ways to form a k-subset in $A_1 \cup \ldots \cup A_k$ than just transversals and subsets of some A_i. This motivates the following definition.

Let $k \geq 2$ and let $\chi_k : [n]^k \to \{-1, +1\}$ be a coloring and $(A_i)_{i < j}$, for some $j \leq k$, be a family of pairwise disjoint subsets of n. Then we define

$$\chi_k(A_0, \ldots, A_{j-1}) = \sum \chi_k(A),$$

where the summation is taken over all sets $A \in [n]^k$ satisfying $A \subseteq \bigcup_{i<j} A_i$ and $A \cap A_i \neq \emptyset$ for every $i < j$. In particular, for $j = k$, the summation goes over all transversals of $A_0, \ldots, A_{k-1}$.

In the following assume that $k \geq 2$ and a coloring $\chi_k : [n]^k \to \{-1, +1\}$ is fixed and let n_0 be the constant from Corollary 10.4. First we show:

Lemma 10.6. *There exists $\epsilon > 0$ such that for every $\chi_k : [n]^k \to \{-1, +1\}$ and for every family $(A_i)_{i<k}$ of pairwise disjoint subsets of n satisfying $|A_0| = \ldots = |A_{k-1}| = t$ for some $t \geq n_0$ there exist $B_0 \subseteq A_0, \ldots, B_{k-1} \subseteq A_{k-1}$ so that*

$$|\chi_k(B_0, \ldots, B_{k-1})| \geq \epsilon t^{\frac{k+1}{2}}.$$

To prove Lemma 10.6 we show

Lemma 10.7. *There exist positive constants $c_1, \ldots, c_{k-1}$ and $d_1, \ldots, d_{k-1}$ so that for every positive integer $j < k$ and every family $(A_i)_{i<j}$ of pairwise disjoint subsets of n satisfying $|A_0| = \ldots = |A_{j-1}| = t$, for some $t \geq n_0$, and for every $\chi_j : [n]^j \to \{-1, +1\}$ we have*

$$|\{(C_0, \ldots, C_{j-1}) \mid \forall i < j : C_i \subseteq A_i \text{ and } |\chi_i(C_0, \ldots, C_{j-1})| \geq c \cdot t^{j/2}\}| \geq d_j \cdot 2^{t \cdot j}.$$

We mimic the argument used to prove the first assertion in the graph-case to show how Lemma 10.7 implies Lemma 10.6.

Proof of Lemma 10.6. Fix some $a \in A_{k-1}$. Then by Lemma 10.7 (for $j = k - 1$ and defining χ_{k-1} by $\chi_{k-1}(C_0, \ldots, C_{k-2}) = \chi_k(C_0, \ldots, C_{k-2}, \{a\})$) we have that

$$|\{(C_0, \ldots, C_{k-2}) \mid |\chi_k(C_0, \ldots, C_{k-2}, \{a\})| \geq c_{k-1} t^{\frac{k-1}{2}}\}| \geq d_{k-1} 2^{t(k-1)}.$$

Put $\delta = \frac{d_{k-1}}{2}$. Then we get the existence of a family $(B_i)_{i<k-1}$, where $B_i \subseteq A_i$, so that

$$|\{a \in A_{k-1} \mid |\chi_k(B_0, \ldots, B_{k-2}, \{a\})| \geq c_{k-1} t^{\frac{k-1}{2}}\}| \geq 2\delta t.$$

Again by symmetry we can assume that

$$|\{a \in A_{k-1} \mid \chi_k(B_0, \ldots, B_{k-2}, \{a\}) \geq c_{k-1} t^{\frac{k-1}{2}}\}| \geq \delta t.$$

Let $B_{k-1} = \{a \in A_{k-1} \mid \chi_k(B_0, \ldots, B_{k-2}, \{a\}) \geq c_{k-1} t^{\frac{k-1}{2}}\}$. Then

$$\chi_k(B_0, \ldots, B_{k-1}) = \sum_{a \in B_{k-1}} \chi_k(B_0, \ldots, B_{k-2}, \{a\}) \geq \delta t \cdot c_{k-1} t^{\frac{k-1}{2}} = \epsilon t^{\frac{k+1}{2}},$$

choosing $\epsilon = \delta \cdot c_{k-1}$, thus proving Lemma 10.6. $\qquad\qquad \square$

Proof of Lemma 10.7. We proceed by induction on j. Observe in the case $j = 1$ we are given a function χ_1 that assigns values to points and we are interested in certain subsets of A_0. This is exactly the situation of Corollary 10.4. The base case of the induction thus follows from Corollary 10.4 (applied for $c = 1$) by choosing $c_1 = 1/2$ and $d_1 = \frac{1}{5}$.

So assume the validity of Lemma 10.7 for some $j \in [1, k-2]$ and fix some $\chi_{j+1} :$ $[n]^{j+1} \to \{-1, +1\}$. Note that for every fixed $a \in A_j$ the function χ_{j+1} naturally gives rise to a function $\chi_j : [n]^j \to \{-1, +1\}$ via $\chi_j(X) := \chi_{j+1}(X \cup \{a\})$. To these function we can then apply the induction hypothesis. Let

$$\mathcal{M} = \{(C_0, \ldots, C_{j-1}, \{a\}) \mid$$

$$C_i \subseteq A_i, i < j, a \in A_j \text{ s.t. } |\chi_{j+1}(C_0, \ldots, C_{j-1}, \{a\})| \geq c_j t^{j/2}\}.$$

Then the induction hypothesis implies that we have for every a at least $d_j 2^{tj}$ subsets $(C_0, \ldots, C_{j-1})$ so that $(C_0, \ldots, C_{j-1}, \{a\}) \in \mathcal{M}$. Thus, we know

$$|\mathcal{M}| \geq t \cdot d_j 2^{tj}.$$

On the other hand we have:

$$|\mathcal{M}| = \sum_{\substack{(C_0, \ldots, C_{j-1}) \\ C_i \subseteq A_i}} |\{a \in A_j \mid (C_0, \ldots, C_{j-1}, \{a\}) \in \mathcal{M}\}|.$$

Here we have 2^{tj} summands, each of which has size (at most) t, that together sum up to at least $d_j t 2^{tj}$. An easy calculation thus gives: there are at least $\frac{d_j}{2} 2^{tj}$ summands which are larger than $\frac{d_j}{2} t$.

Fix such a $(C_0, \ldots, C_{j-1})$. Then there are $\frac{d_j}{2} t$ many $a \in A_j$ such that $(C_0, \ldots, C_{j-1}, \{a\}) \in \mathcal{M}$ meaning that $|\chi_{j+1}(C_0, \ldots, C_{j-1}, \{a\})| \geq c_j t^{j/2}$. If we thus let

$$x_a = \frac{1}{c_j} t^{-j/2} \chi_{j+1}(C_0, \ldots, C_{j-1}, \{a\}).$$

for every $a \in A_j$, then $|x_a| \geq 1$ for at least $\frac{d_j}{2} t$ many $a \in A_j$.

Apply Corollary 10.4 with respect to $c = \frac{d_j}{2}$. Then we have for at least $\frac{1}{5} 2^t$ choices $C_j \subseteq A_j$ that

$$|\chi_{j+1}(C_0, \ldots, C_j)| = \left| \sum_{a \in C_j} \chi_{j+1}(C_0, \ldots, C_{j-1}, \{a\}) \right| = c_j t^{j/2} \left| \sum_{a \in C_j} x_a \right|$$

$$\geq c_j t^{j/2} \cdot \frac{d_j}{4} t^{1/2} = \frac{c_j d_j}{4} t^{(j+1)/2}.$$

As this is true for at least $\frac{d_j}{2} 2^{tj}$ choices of $(C_0, \dots, C_{j-1})$, choosing $c_{j+1} = \frac{c_j d_j}{4}$ and $d_{j+1} = \frac{d_j}{10}$ completes the proof of Lemma 10.7. $\qquad\square$

In the next step we transform the imbalance of a product into an imbalance for a set:

Lemma 10.8. *Let $\eta > 0$ and $j \leq k$ be a positive integer. Then there exists an $\xi = \xi(\eta, j) > 0$ such that for every $\chi_k : [n]^k \to \{-1, +1\}$ and every family $(B_i)_{i<j}$ of pairwise disjoint subsets of n, we have that $|\chi_k(B_0, \dots, B_{j-1})| \geq \eta n^{\frac{k+1}{2}}$ implies the existence of some $I \subseteq j$ satisfying*

$$|\chi_k(\bigcup_{i \in I} B_i)| \geq \xi n^{\frac{k+1}{2}}.$$

Proof. For $j = 1$ and every $\eta > 0$ the lemma is trivial choosing $\xi = \eta$. So assume the validity of Lemma 10.8 for some $j < k$ and all $\eta > 0$, and let $(B_i)_{i \leq j}$ be a family of pairwise disjoint subsets of n such that $|\chi_k(B_0, \dots, B_j)| \geq \eta n^{\frac{k+1}{2}}$, for some $\eta > 0$.

Observe that

$$\chi_k(B_0, \dots, B_j) = \chi_k(\bigcup_{i \leq j} B_i) - \sum \chi_k(B_{i_1}, \dots, B_{i_\ell}),$$

where the summation is taken over all proper (and nonempty) subfamilies of $B_0, \dots, B_j$. Hence, at least one of the summands of the right hand side has absolute value at least $\frac{\eta}{2^{j+1}} n^{\frac{k+1}{2}}$. If $\chi_k(\bigcup_{i \leq j} B_i)$ has this size, we are done. Otherwise, applying the inductive hypothesis to the appropriate summand replacing η by $\frac{\eta}{2^{j+1}}$ proves Lemma 10.8. $\qquad\square$

Now the proof of Theorem 10.1 is easily finished. Without loss of generality we can assume that $n = k \cdot t$. Let $\chi_k : [n]^k \to \{-1, +1\}$ be a coloring and $A_0 \cup \dots \cup A_{k-1} = n$ be a partition of n into k disjoint sets each of size $t \geq n_0$. Then, by Lemma 10.6, there exist $B_0 \subseteq A_0, \dots, B_{k-1} \subseteq A_{k-1}$ so that

$$|\chi_k(B_0, \dots, B_{k-1})| \geq \epsilon t^{\frac{k+1}{2}} = \epsilon k^{-\frac{k+1}{2}} n^{\frac{k+1}{2}}.$$

Applying Lemma 10.8 for $\eta = \epsilon k^{-\frac{k+1}{2}}$ and $j = k$ yields a constant $\xi = \xi(\eta, k)$ and a subset $I \subseteq k$ satisfying

$$|\chi_k(\bigcup_{i \in I} B_i)| \geq \xi n^{\frac{k+1}{2}}.$$

Choosing $c_0 \leq \xi$ in such a way that c_0 takes care of all small n completes the proof of Theorem 10.1. $\qquad\square$

Chapter 11
Partition Relations for Cardinal Numbers

Recall the infinite version of Ramsey's theorem: $\omega \rightarrow (\omega)_r^k$, whenever k, r are positive integers. The aim of this section is to discuss some extensions of this relation to larger cardinals. Our treatment will be far from complete. For ω more results on this topic we refer the reader to the book of Erdős et al. (1984).

We start with a negative result, proved by Erdős and Rado (1952) which shows that the exponent k may not be replaced by an infinite cardinal without conflicting with the axiom of choice.

Proposition 11.1. *Let $\kappa \geq \omega$ be a cardinal. Then*

$$\kappa \nrightarrow (\omega)_2^\omega,$$

where $\nrightarrow$ denotes the negation of $\rightarrow$.

Proof. Let $<_{\text{well}}$ be a well-ordering of $[\kappa]^\omega$, the set of countable subsets of κ. We define a coloring $\Delta : [\kappa]^\omega \rightarrow 2$ witnessing to $\kappa \nrightarrow (\omega)_2^\omega$ as follows:

$$\Delta(A) = \begin{cases} 0, & \text{if there exists } B \subset A \text{ such that } B <_{\text{well}} A \\ 1, & \text{otherwise.} \end{cases}$$

Now let $F \in [\kappa]^\omega$ and let $A = \{a_i \mid i < \omega\} \in [F]^\omega$ be the first ω-subset with respect to $<_{\text{well}}$ in F. Take any proper ω-subset $B \subset A$, then $A <_{\text{well}} B$ and therefore $\Delta(A) = 1$. On the other hand, let $A^* = \{a_{2i+1} \mid i < \omega\}$ and for each $m < \omega$ let $A_m = \{a_0, a_2, \ldots, a_{2m}\} \cup A^*$. Put $A_{m_0} = \min\{A_m \mid m < \omega\}$, where the minimum is taken with respect to the well-ordering $<_{\text{well}}$. Then $A_{m_0} <_{\text{well}} A_{m_0+1}$ and $A_{m_0} \subset A_{m_0+1}$. Hence, $\Delta(A_{m_0+1}) = 0$ which proves Proposition 11.1. $\qquad\square$

This result prevents us from considering colorings of infinite subsets in this chapter. But observe that the proof given above uses essentially the Axiom of Choice, i.e., Zermelo's well-ordering theorem. If one drops the Axiom of Choice, even the relation $\omega \rightarrow (\omega)_2^\omega$ may be consistent, cf., e.g., Mathias (1969) and Kleinberg (1970).

H.J. Prömel, *Ramsey Theory for Discrete Structures*,
DOI 10.1007/978-3-319-01315-2_11,
© Springer International Publishing Switzerland 2013

Throughout this chapter we assume the Axiom of Choice. All set-theoretic notions used are standard and can be found, e.g., in Jech (1978).

Following the convention introduced by John von Neumann we identify ordinals with the set of their predecessors and cardinals with their initial ordinals. For every cardinal κ let κ^+ denote the least cardinal greater than κ, i.e., the (cardinal) successor of κ. A cardinal κ is called regular, if for every $\lambda < \kappa$ and any choice of subsets $A_\nu \subseteq \kappa$ for $\nu < \lambda$ with $|A_\nu| < \kappa$ it follows that $|\bigcup_{\nu<\lambda} A_\nu| < \kappa$, in other words, κ cannot be written as the union of less than κ many sets of cardinality less than κ. It can easily be shown, using the Axiom of Choice, that every successor cardinal is regular.

Addition and multiplication of infinite cardinals κ and λ is easy:

$$\kappa + \lambda = \kappa \cdot \lambda = \max\{\kappa, \lambda\}.$$

Exponentiation, in general, is more difficult, but for our purposes it is enough to know that $\lambda \leq \kappa$ implies that $\lambda^\kappa = 2^\kappa$ and $\lambda^n = \lambda$ for every finite n.

As usual we denote the first infinite cardinal also by $\aleph_0$, i.e., $\aleph_0 = \omega$, and the second one, the first uncountable cardinal, by $\aleph_1$.

Section 11.1 is devoted to the proof of the Erdős-Rado partition theorem for cardinals, in Sect. 11.2 some negative partition relations are given essentially showing that the Erdős-Rado theorem is best possible in the sense that the Ramsey numbers are correctly estimated. In Sect. 11.3 Dushnik-Miller's theorem (for regular cardinals) is discussed. In Sect. 11.4 we consider the question for which cardinals κ other than ω the relation $\kappa \to (\kappa)^2_2$ might be true. Finally, in Sect. 11.5 we glance briefly at canonical partition relations for cardinals.

11.1   Erdős-Rado's Partition Theorem for Cardinals

The following quite general partition relation for cardinals is due to Erdős and Rado (1956). Let $\exp_0(\kappa) = \kappa$ and $\exp_{k+1}(\kappa) = 2^{\exp_k(\kappa)}$.

Theorem 11.2 (Erdős, Rado). *Let $\kappa \geq \omega$ be a cardinal and k be a positive integer. Then*

$$\exp_{k-1}(\kappa)^+ \to (\kappa^+)^k_\kappa.$$

Proof. We proceed by induction on k, the case $k = 1$, i.e., $\kappa^+ \to (\kappa^+)^1_\kappa$, reduces to the pigeonhole principle. So assume that the theorem is valid for some $k \geq 1$, put $\lambda = \exp_{k-1}(\kappa)$ and let $\Delta : \binom{(2^\lambda)^+}{k+1} \to \kappa$ be a coloring. We want to find a monochromatic set $F \subset (2^\lambda)^+$ of size κ^+. For each $x < (2^\lambda)^+$ let $\Delta_x : \binom{(2^\lambda)^+ \setminus \{x\}}{k} \to \kappa$ be defined by $\Delta_x(A) := \Delta(A \cup \{x\})$. We claim:

1. There exists a set $S \subset (2^\lambda)^+$ of cardinality $|S| = 2^\lambda$ such that for every $M \subset S$ with $|M| \leq \lambda$ and every $x \in (2^\lambda)^+ \setminus S$ there exists $y = y(M, x) \in S \setminus M$ so that $\Delta_x \rceil [M]^k = \Delta_y \rceil [M]^k$.

Before proving (1) we show how this implies the theorem. Fix some $x \in (2^\lambda)^+ \setminus S$. By transfinite induction we define a set $Y = \{y_\mu \mid \mu < \lambda^+\} \subseteq S$ as follows. Let $y_0 \in S$ be arbitrary and assume that $\{y_\mu \mid \mu < \nu\} = M$ has been defined for some $\nu < \lambda^+$. Then let $y_\nu = y(M, x)$ be according to (1). Observe that for every $A \in \binom{Y}{k+1}$, where $A = \{y_{i_0}, \ldots, y_{i_{k-1}}, y_\nu\}$ with $i_0 < \ldots < i_{k-1} < \nu$, we have that

$$\Delta(\{y_{i_0}, \ldots, y_{i_{k-1}}, y_\nu\}) = \Delta_{y_\nu}(\{y_{i_0}, \ldots, y_{i_{k-1}}\}) = \Delta_x(\{y_{i_0}, \ldots, y_{i_{k-1}}\}). \quad (11.1)$$

Now consider $\Delta_x : [Y]^k \to \kappa$. Since $|Y| = \lambda^+$ and according to the inductive hypothesis there exists an $F \subseteq Y$ with $|F| = \kappa^+$ so that $\Delta_x(A) = \Delta_x(B)$ for all $A, B \in [F]^k$. Thus, the theorem follows from (11.1).

It remains to prove (1). We define an ascending sequence $S_0 \subseteq S_1 \subseteq \ldots \subseteq S_\mu \subseteq \ldots, \mu < \lambda$, of subsets of $(2^\lambda)^+$, each of size 2^λ, as follows.

Choose $S_0 \subset (2^\lambda)^+$ with $|S_0| = 2^\lambda$ arbitrarily and for each limit ordinal ν let $S_\nu = \bigcup_{\mu < \nu} S_\mu$. Now assume that S_μ with $|S_\mu| = 2^\lambda$ has been defined. We now define $S_{\mu+1}$.

Observe that there exist at most $(2^\lambda)^\lambda = 2^\lambda$ subsets of S_μ of size λ and therefore there exist at most $\lambda \cdot 2^\lambda = 2^\lambda$ subsets M of S_μ of size at most λ. Fix such an $M \subseteq S_\mu$. Then there exist at most 2^λ mappings $f : [M]^k \to \kappa$ (recall that $\kappa \leq \lambda$). This shows that

$$|\{\Delta_x \rceil [M]^k \mid x \in (2^\lambda)^+ \setminus M\}| \leq 2^\lambda.$$

Assume a well-ordering on $(2^\lambda)^+$ to be given and for every $x \in (2^\lambda)^+ \setminus S_\mu$ let $y(M, x)$ be the smallest $y \in (2^\lambda)^+ \setminus M$ such that $\Delta_x \rceil [M]^k = \Delta_y \rceil [M]^k$. Denote the set of those y by $Y(M)$. Then $|Y(M)| \leq 2^\lambda$. Now put

$$S_{\mu+1} = S_\mu \cup \bigcup Y(M),$$

where the union is taken over all $M \subseteq S_\mu$ with $|M| \leq \lambda$. Then $|S_{\mu+1}| \leq 2^\lambda + 2^\lambda \cdot 2^\lambda = 2^\lambda$. Finally let $S = \bigcup_{\mu < \lambda} S_\mu$. Then $|S| \leq \lambda^+ \cdot 2^\lambda = 2^\lambda$ and, by construction, S has the desired properties. $\square$

It seems to be worth while to state the following special case explicitly:

Corollary 11.3. *For every $\kappa \geq \omega$ it follows that $(2^\kappa)^+ \to (\kappa^+)^2_\kappa$, and, even more special, $(2^{\aleph_0})^+ \to (\aleph_1)^2_{\aleph_0}$.* $\square$

11.2 Negative Partition Relations

Next we are going to show that Theorem 11.2 is, in a sense, best possible. Before we will do this in general, we prove that Corollary 11.3 is the best we can expect.

Let κ be a cardinal. Then we denote by 2^κ the set of all sequences of length κ over the alphabet $2 = \{0, 1\}$. Hence, every $x \in 2^\kappa$ can be written as $x = (x(0), \dots, x(\nu), \dots)$ where $x(\nu) < 2$ for every $\nu < \kappa$. The natural order on 2, i.e., $0 < 1$, gives a lexicographic order on 2^κ which will be denoted by $\prec$. So $x \prec y$ if and only if $x(\nu) < y(\nu)$ where ν is the least ν such that $x(\nu) \neq y(\nu)$. In fact, we know that then $x(\nu) = 0$ and $y(\nu) = 1$.

Lemma 11.4. *For every $\kappa \geq \omega$ it follows that $2^\kappa \nrightarrow (3)^2_\kappa$, and hence $2^{\aleph_0} \nrightarrow (3)^2_{\aleph_0}$.*

Proof. Let $\Delta : [2^\kappa]^2 \to \kappa$ be defined by $\Delta(\{x, y\})$ being the least position $\nu < \kappa$ such that $x(\nu) \neq y(\nu)$. Obviously it is impossible to have pairwise distinct $x, y, z \in 2^\kappa$, such that $\Delta(\{x, y\}) = \Delta(\{x, z\}) = \Delta(\{y, z\})$.   □

The following result of Sierpiński (1933) shows in particular that the straightforward generalization of Ramsey's theorem, viz. $\aleph_1 \to (\aleph_1)^2_2$, is false.

Theorem 11.5 (Sierpiński). *For every $\kappa \geq \omega$ it follows that $2^\kappa \nrightarrow (\kappa^+)^2_2$, and hence $2^{\aleph_0} \nrightarrow (\aleph_1)^2_2$.*

Proof. We will derive Theorem 11.5 from the following fact:

1. There does not exist any increasing or decreasing κ^+-sequence in 2^κ with respect to $\prec$.

We show that 2^κ has no increasing κ^+-sequence. The decreasing case can be handled analogously. To derive a contradiction assume that $X = \{x_\nu : \nu < \kappa^+\} \subseteq 2^\kappa$ is an increasing κ^+-sequence, i.e., $x_\mu \prec x_\nu$ whenever $\mu < \nu$. For each $\nu < \kappa^+$ and each $\mu < \kappa$ let $x_\nu {\upharpoonright} \mu = (x_\nu(0), \dots, x_\nu(\mu'), \dots)$, $\mu' < \mu$ be the initial segment of length μ of x_ν. Now let $\eta \leq \kappa$ be the least ordinal such that $|\{x_\nu {\upharpoonright} \eta \mid \nu < \kappa^+\}| = \kappa^+$. Without loss of generality we can assume that $x_\mu {\upharpoonright} \eta \neq x_\nu {\upharpoonright} \eta$ for all x_μ and x_ν in X. Otherwise one could choose an appropriate subset of X which is still of size κ^+. Define a sequence $d_\nu, \nu < \kappa^+$, where d_ν gives the least position at which x_ν and $x_{\nu+1}$ differ. By our assumption on X we know that $d_\nu < \eta \leq \kappa$ for every $\nu < \kappa^+$. Thus there exists $\gamma < \eta$ such that $d_\nu = \gamma$ for κ^+ many ν. Observe that $|\{x_\nu {\upharpoonright} \gamma \mid \nu < \kappa^+\}| \leq \kappa$. So let $\mu' < \mu$ be such that $d_{\mu'} = d_\mu = \gamma$ and $x_{\mu'} {\upharpoonright} \gamma = x_\mu {\upharpoonright} \gamma$. Then $x_\mu \prec x_{\mu'+1}$. But on the other hand, since $\mu' < \mu' + 1 \leq \mu$, we have $x_{\mu'+1} \preceq x_\mu$, a contradiction which proves (1).

We now prove Theorem 11.5 by defining a 2-coloring of $[2^\kappa]^2$ which does not admit a monochromatic κ^+-set. So let $x_\nu, \nu < 2^\kappa$, be any enumeration of 2^κ and let $\Delta : [2^\kappa]^2 \to 2$ be given by

$$\Delta(x_\mu, x_\nu) = \begin{cases} 0, & \text{if } \mu < \nu \text{ and } x_\mu \prec x_\nu \\ 1, & \text{if } \mu < \nu \text{ and } x_\nu \prec x_\mu. \end{cases}$$

Then a monochromatic set of size κ^+ would contradict (1).   □

The following more general results are contained in Erdős et al. (1965).

Theorem 11.6. *Let $\kappa \geq \omega$ be a cardinal and $k \geq 2$ be a positive integer. Then*

$$\exp_{k-1}(\kappa) \nrightarrow (\kappa^+)_2^k.$$

Theorem 11.7. *Let $\kappa \geq \omega$ be a cardinal and $k \geq 3$ be a positive integer. Then*

$$\exp_{k-1}(\kappa) \nrightarrow (k+1)_\kappa^k.$$

11.3   Dushnik-Miller's Theorem

By Theorem 11.6 we have that $\aleph_1 \nrightarrow (\aleph_1)_2^2$. On the other hand, Ramsey's theorem trivially implies $\aleph_1 \rightarrow (\omega)_2^2$. In this section we prove a partition relation which is, in a sense, halfway between these two relations.

For cardinals κ, λ_0 and λ_1 let $\kappa \rightarrow (\lambda_0, \lambda_1)_2^2$ denote the assertion that for every 2-coloring of $[\kappa]^2$ there exists either a set of size λ_0 which is monochromatic in color 0 or a set of size λ_1 which is monochromatic in color 1. Hence, in particular, $\aleph_1 \rightarrow (\omega, \omega)_2^2$. The following result is due to Dushnik and Miller (1941):

Theorem 11.8. *Let $\kappa \geq \omega$ be a regular cardinal. Then $\kappa \rightarrow (\kappa, \omega)_2^2$ and, in particular, $\aleph_1 \rightarrow (\aleph_1, \omega)_2^2$.*

We should mention that Theorem 11.8 is also true for singular (i.e., non-regular) cardinals, cf. Dushnik and Miller (1941).

Proof of Theorem 11.8. Let $\Delta : [\kappa]^2 \rightarrow 2$ be a coloring. First, we show:

1. If for every $S \subseteq \kappa$ of size κ there exists an $x \in S$ such that $|\{y \in S \mid \Delta(\{x, y\}) = 1\}| = \kappa$ then there exists a countable set $D \subseteq \kappa$ such that D is monochromatic with color 1.

Let $\Gamma(x) = \{y < \kappa \mid \Delta(\{x, y\}) = 1\}$. Choose $d_0 < \kappa$ arbitrarily such that $|\Gamma(d_0)| = \kappa$. Now assume that $D_n = \{d_0, \ldots, d_n\}$ is defined such that $\Delta\!\restriction\![D_n]^2 \equiv 1$ and such that $|S_n| = \kappa$, where $S_n = \bigcap\{\Gamma(d_i) \mid d_i \in D_n\}$. Then choose $d_{n+1} \in S_n$ such that $|\{y \in S_n \mid \Delta(\{d_{n+1}, y\}) = 1\}| = \kappa$. Clearly, $D = \bigcup_{n<\omega} D_n$ satisfies (1).

So we assume that there is no countable set $D \subseteq \kappa$ with $\Delta\!\restriction\![D]^2 \equiv 1$. Then let $S \subseteq \kappa$ be of size κ so that for every element $x \in S$ if follows that $|\{y \in S \mid \Delta(\{x, y\}) = 1\}| < \kappa$. We construct recursively a sequence x_ν, such that $\Delta(\{x_\mu, x_\nu\}) = 0$ whenever $\mu < \nu < \kappa$. Assume that $(x_\nu \in S \mid \nu < \nu')$ have been constructed for some $\nu' < \kappa$. Then $|S \cap (\bigcup_{\nu<\nu'} \Gamma(x_\nu))| < \kappa$. Notice that here the regularity of κ is needed. Now choose $x_{\nu'} \in S \setminus \bigcup_{\nu<\nu'} \Gamma(x_\nu)$, completing the proof of Theorem 11.8.   □

11.4　Weakly Compact Cardinals

As shown in Theorem 11.2, for every pair λ, ρ of cardinals, where $\rho < \lambda$, and every positive integer k there exists a cardinal κ such that $\kappa \to (\lambda)^k_\rho$. Moreover, in case λ is a successor cardinal we have determined the smallest κ satisfying this relation. But it seems to be a natural question to ask whether κ can be λ, in particular whether the relation $\kappa \to (\kappa)^2_2$ can hold for cardinals other than $\kappa = \omega$. The answer to this question leads immediately to large cardinals.

Let κ be an uncountable cardinal, i.e., $\kappa > \omega$. Then κ is called inaccessible if κ is regular and $2^\lambda < \kappa$ for every $\lambda < \kappa$. Inaccessible cardinals were introduced by Sierpiński and Tarski (1930). In particular they have the property that $|X| < \kappa$ implies $|\mathcal{P}(X)| < \kappa$. This and some other properties inaccessible cardinals share with ω. So, in a sense, one can say that an inaccessible cardinal is related to smaller cardinals as ω is related to finite cardinals. But it is not at all clear whether inaccessible cardinals do exist. To be more precise: One can show that the existence of such cardinals cannot be proved in ZF + Axiom of Choice. Erdős et al. (1965) showed that the requirement $\kappa \to (\kappa)^2_2$ leads at least to inaccessible cardinals.

Theorem 11.9. *If $\kappa > \omega$ and $\kappa \to (\kappa)^2_2$, then κ is inaccessible.*

Proof. We have to show that κ is regular and that $2^\lambda < \kappa$ for every $\lambda < \kappa$. The second assertion follows immediately from Sierpiński's Theorem 11.5. Assume that $\kappa \le 2^\lambda$ for some $\lambda < \kappa$. Then $2^\lambda \not\to (\lambda^+)^2_2$ implies $\kappa \not\to (\lambda^+)^2_2$ and hence $\kappa \not\to (\kappa)^2_2$.

So it remains to show that κ is regular. Suppose not. Then there exists a family $X_\nu, \nu < \lambda$, for some $\lambda < \kappa$ of pairwise disjoint sets such that $|X_\nu| < \kappa$ for each $\nu < \lambda$ and $\kappa = |\bigcup\{X_\nu \mid \nu < \lambda\}|$. Define $\Delta : [\kappa]^2 \to 2$ by $\Delta(\{x, y\}) = 0$ if $\{x, y\} \subseteq X_\nu$ for some $\nu < \lambda$, $\Delta(\{x, y\}) = 1$, otherwise. Obviously, there does not exist $M \in [\kappa]^\kappa$ which is monochromatic with respect to Δ, thus contradicting $\kappa \to (\kappa)^2_2$. $\square$

Cardinals $\kappa > \omega$ satisfying $\kappa \to (\kappa)^2_2$ are called weakly compact. From what is said before it follows that their existence cannot be proved in ZF + Axiom of Choice. In a sense, the situation is even worse. One can show that $\kappa \to (\kappa)^2_2$ fails for many inaccessible cardinals including the first one provided such numbers exist at all. Moreover, even if the existence of an inaccessible cardinal is assumed it cannot be proved in ZF + Axiom of Choice that there is a weakly compact cardinal. For a detailed discussion and an extensive bibliography on this topic, compare Erdős et al. (1984).

We close this paragraph with stating a result which shows that if there exists a weakly compact cardinal it has indeed quite strong partition properties.

Theorem 11.10. *If $\kappa \to (\kappa)^2_2$, then $\kappa \to (\kappa)^k_\rho$, for every $k < \omega$ and every $\rho < \kappa$.*

This result can be shown using similar arguments as in the proof of Ramsey's theorem. We omit the proof.

11.5   Canonical Partition Relations for Cardinals

Finally, we briefly review some canonical partition results for infinite cardinals. The canonical Ramsey arrow extends naturally to arbitrary cardinals, $\kappa \overset{can}{\rightarrow} (\lambda)^k$ meaning that for every coloring Δ of the k-subsets of κ with arbitrary many colors there exists a λ-subset $F \in [\kappa]^\lambda$ of κ so that $\Delta \rceil [F]^k$ is canonical. For $\kappa = \lambda = \omega, k < \omega$, this relation was shown in the Erdős-Rado canonization theorem (Theorem 1.4). The argument given there to prove $\omega \overset{can}{\rightarrow} (\omega)^k$ actually shows that if λ, κ and k are cardinals with $\lambda > 2k$ such $\kappa \rightarrow (\lambda)^{2k}_k$ then $\kappa \overset{can}{\rightarrow} (\lambda)^k$. Combining this observation with Theorem 11.10 we obtain immediately:

Theorem 11.11. *If $\kappa \rightarrow (\kappa)^2_2$, then $\kappa \overset{can}{\rightarrow} (\kappa)^k$ for every $k < \omega$.*   □

Moreover, applying the relation $\exp_{k-1}(\kappa)^+ \rightarrow (\kappa^+)^k_k$ (instead of Ramsey's theorem) in the proof of Theorem 1.4 yields that

$$\exp_{2k-1}(\kappa)^+ \overset{can}{\rightarrow} (\kappa^+)^\kappa.$$

However, this is far from best possible. Baumgartner (1975) showed that the same cardinal which satisfies the Erdős-Rado partition relation is already large enough for the canonical partition relation:

Theorem 11.12. *Let $\kappa \geq \omega$ be a cardinal and k be a positive integer. Then*

$$\exp_{k-1}(\kappa)^+ \overset{can}{\rightarrow} (\kappa^+)^k.$$

We omit the proof of this result which combines ideas behind a theorem of Fodor (1956) on regressive mappings and the Erdős-Rado canonization theorem.

Part IV
Graphs and Hypergraphs

Chapter 12
Finite Graphs

Although our graph theoretic terminology is standard, let us briefly recall the basic definitions, compare also any standard book on graph theory, e.g. Bollobás (1998) and Diestel (2010).

A graph is an ordered pair $G = (V, E)$, where V is the set of vertices of G and $E \subseteq [V]^2$ is the set of edges of G. So edges are two-element subsets of vertices and graphs in our sense are simple graphs without multiple edges and without loops. The vertex set of a graph G is denoted by $V(G)$, its cardinality by $v_G := |V(G)|$. Similarly, the edge set of G is denoted by $E(G)$ and its cardinality by $e_G := |E(G)|$.

A graph is called finite if its vertex set is finite.

A subgraph of G is given by a subset $W \subseteq V$ of vertices together with the edge set $E_W \subseteq E \cap [W]^2$. If $E_W = E \cap [W]^2$ then it is called an *induced* subgraph, as the subgraph is induced from its vertex set. The subgraph of G induced by W is denoted by $G[W]$.

Two graphs $G = (V, E)$ and $G' = (V', E')$ are isomorphic if there exists a bijection $f : V \to V'$ such that $\{x, y\} \in E$ if and only if $\{f(x), f(y)\} \in E'$ for all $x, y \in V$. Such a mapping f is a graph *isomorphism* between G and G'.

We use the Ramsey arrow $G \to (F)_r^H$ as a shorthand abbreviation for the following Ramsey-type statement: for every r-coloring of the H-subgraphs of G (i.e., of all subgraphs of G which are isomorphic to H) there exists a F-subgraph with all its H-subgraphs in the same color. If $H = K_1$ (i.e, we color vertices), we also use the notation $G \to (F)_r^v$. Similarly, in the case of edges ($H = K_2$) we use $G \to (F)_r^e$. We use the notation $G \overset{\text{ind}}{\to} (F)_r^H$ if for every r-coloring of the *induced* H-subgraphs of G there exists an induced F-subgraph with all its induced H-subgraphs in the same color. Note that if F does not contain any H-copy then we trivially have $G \to (F)_r^H$ and $G \overset{\text{ind}}{\to} (F)_r^H$ whenever G contains an (induced) F-subgraph.

Note that for $H = K_\ell$, for some $\ell \geq 1$, $G \overset{\text{ind}}{\to} (F)_r^H$ implies $G \to (F)_r^H$ and $G \not\to (F)_r^H$ implies $G \overset{\text{ind}}{\not\to} (F)_r^H$. We will thus mostly be concerned with proving positive statements in the induced setting and negative statements in the weak setting.

H.J. Prömel, *Ramsey Theory for Discrete Structures*,
DOI 10.1007/978-3-319-01315-2_12,
© Springer International Publishing Switzerland 2013

A fundamental question in graph Ramsey theory is: given graphs F and H what properties should a graph G have in order to guarantee that $G \to (F)_r^H$ resp. $G \overset{ind}{\to} (F)_r^H$. We first discuss this question for $H = K_1$, i.e., vertex-colorings.

12.1   Vertex Colorings

Recall that Ramsey's Theorem implies that for every m, r there exists an $n = n(m, r)$ such that $K_n \to (K_m)_r^v$. This trivially also implies that for every F and every r there exists an $n = n(G, r)$ such that $K_n \to (F)_r^v$. The following theorem shows that it is also possible to replace the (weak) Ramsey arrow by the induced one.

Theorem 12.1. *For every finite graph F and $r \geq 1$ there exists a graph G such that $G \overset{ind}{\to} (F)_r^v$.*

Proof. Let $G = F^r$ be the r-th power of F, i.e., for $F = (V, E)$ define a graph on V^r by joining $(x_0, \ldots, x_{r-1})$ and $(y_0, \ldots, y_{r-1})$ by an edge if there exists $j < r$ such that $x_i = y_i$ for all $i < j$ and $\{x_j, y_j\} \in E$. Figure 12.1 shows the second power of C_4, a cycle on four vertices.

Now $G \overset{ind}{\to} (F)_r^v$ can be seen as follows. If for every $v \in V$ there exists an r-tuple $(v, x_2, \ldots, x_r) \in V^r$ colored, say, with red, then the graph induced by one such r-tuple for each $v \in V$ induces a red F-subgraph. Otherwise, there exists $v \in V$ such that all r-tuples $(v, x_2, \ldots, x_r)$ are colored with $r - 1$ colors. By fixing the first coordinate to be v, we can repeat the same argument on the remaining coordinates, thus eventually reaching the situation where we are left with only one color in which case a monochromatic F is inevitable. $\square$

One of the main questions of graph Ramsey theory is to impose additional restrictions on the graph G – and then ask whether under these restrictions a graph G with $G \overset{ind}{\to} (F)_r^v$ still exists. Of course, one can always give a trivial restriction of not containing F as a subgraph. The following lemma shows that $G \overset{ind}{\to} (F)_r^v$ doesn't hold even under much weaker conditions. To state it we need some notations.

Notation. For a graph $G = (V, E)$ we denote by $d(G) = e_G/v_G$ the density of G and by $m(G) = \max_{J \subseteq G} \frac{e_J}{v_J}$ the *maximum density* of G, i.e., the density of the densest subgraph of G. Furthermore let $\delta(G)$ denote the minimum degree in G, i.e., $\delta(G) := \min_{v \in V(G)} \deg(v)$, and let $\delta_{max}(G) := \max_{G' \subseteq G} \delta(G')$ be the maximum minimum degree in all subgraphs of G.

Lemma 12.2. *If F and G are two graphs such that $m(G) < \frac{1}{2}r\delta_{max}(F)$ for some $r \geq 2$, then $G \nrightarrow (F)_r^v$.*

Proof. Let $F' \subseteq F$ such that $\delta(F') = \delta_{max}(F)$. We show that for every graph G with $m(G) < \frac{1}{2}r\delta(F')$ we find a vertex coloring of G without a monochromatic F', and hence also without a monochromatic F. Assume this is not true. Then there

Fig. 12.1 The graph $(C_4)^2$

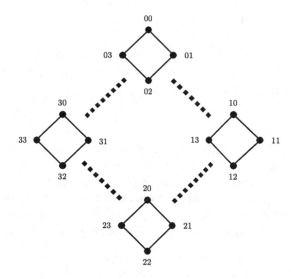

exists a minimal counterexample G_0. As G_0 is minimal we know that for every vertex $v \in V(G_0)$ the graph $G_0 - v$ does have a vertex coloring without a monochromatic F'. Fix an arbitrary vertex $v \in V(G_0)$ and consider any such coloring of $G_0 - v$. Clearly, if $\deg(v) < r\delta(G_0)$ then there exists a color $1 \leq i \leq r$ used by less than $\delta(F')$ neighbors of v. Assigning color i to v we have that there are less than $\delta(G')$ i-colored neighbors of v, thus v cannot belong to a monochromatic copy of F'. Therefore we know that in G_0 every vertex has degree at least $r\delta(F')$, that is $m(G_0) \geq d(G_0) = (\sum_v \deg(v))/(2v_{G_0}) \geq \frac{1}{2}r\delta(F') = \frac{1}{2}r\delta_{max}(F)$. A contradiction. □

Lemma 12.2 states that if G is getting too sparse then it cannot arrow F. However, between this negative result and Theorem 12.1 there is still much room. One of the starting points of graph Ramsey theory is a result of Folkman (1970) who proved a restricted partition theorem. To state it let us denote by $cl(G)$ the size of the largest complete subgraph in G.

Theorem 12.3 (Folkman). *Let r be a positive integer and let F be a finite graph. Then there exists a finite graph G satisfying*

$$cl(G) = cl(F) \qquad and \qquad G \overset{ind}{\to} (F)_r^v.$$

Observe that if the restriction $cl(G) = cl(F)$ is dropped completely then such a result follows from Theorem 12.1. The construction given in that theorem actually provides a graph G such that $cl(G) = cl(F)^r$.

To prove Theorem 12.3 we define a slightly modified way of composing graphs. Let $G = (V_G, E_G)$ and $H = (V_H, E_H)$ be graphs and let $\mathcal{F}$ be the set of mappings from V_G into V_H. The graph $G\{H\} = (W, X)$ is defined as follows:

$$W = V_G \times V_H \times \mathcal{F}$$

$$X = E_0 \cup E_1, \quad \text{where}$$

$$E_0 = \{\{(x, y, f), (x, y', f')\} \mid x \in V_G, y, y' \in V_H, \{y, y'\} \in E_H \text{ and } f, f' \in \mathcal{F}\}$$

$$E_1 = \{\{(x, y, g), (x', y', g)\} \mid x, x' \in V_G, \{x, x'\} \in E_G, y, y' \in V_H, g \in \mathcal{F} \text{ and}$$

$$g(x) = y \text{ and } g(x') = y'\}.$$

Observe that every triangle in $G\{H\}$, and therefore every clique, has its edges either completely in E_0 or completely in E_1. On the other hand, every clique in E_0 corresponds to a clique in H (considering the second coordinate) and every clique in E_1 corresponds to a clique in G (considering the first coordinate). So we have that $cl(G\{H\}) = \max\{cl(G), cl(H)\}$, in particular $cl(G\{G\}) = cl(G)$.

Using this simple observation we prove Folkman's theorem:

Proof of Theorem 12.3. We prove it for $r = 2$, claiming that $F\{F\} \to (F)_2^v$. Let $F = (V_F, E_F)$ and assume the vertices of $F\{F\}$ are two-colored using colors red and blue, say.

Assume there exists a vertex $x \in V_F$ such that for every vertex $y \in V_F$ there exists a mapping $f : V_F \to V_F$ (i.e., $f \in \mathcal{F}$) so that the vertex (x, y, f) is colored red. In this case there exists a red F-subgraph (with all edges in E_0). Otherwise, for every vertex $x \in V_F$ there exists a vertex $y(x) \in V_F$ such that all vertices $(x, y(x), f)$ with $f \in \mathcal{F}$ are colored blue. Consider the mapping $f : V_F \to V_F$ defined by $f(x) = y(x)$. Then $\{(x, y(x), f) \mid x \in V_F\}$ spans a blue F-subgraph (with all edges in E_1). □

This proof is due to Komjáth and Rödl (1986) where an analogous method is used to prove restricted vertex partition theorems for infinite graphs as well.

12.2 Colorings of Edges

Thinking about Ramsey's theorem in terms of graphs, the next step after vertex colorings obviously is asking about edge colorings. As in the vertex case, Ramsey's Theorem implies that for every m and r there exists an $n = n(m, r)$ such that $K_n \to (K_m)_r^e$. The following theorem shows that for every pair F and r there exists an G such that $G \xrightarrow{\text{ind}} (F)_r^e$. For the proof we follow Nešetřil and Rödl (1985).

Theorem 12.4. *Let r be a positive integer and let F be a finite graph. Then there exists a finite graph G satisfying $G \xrightarrow{\text{ind}} (F)_r^e$.*

Proof. Define a graph on $\mathcal{P}(n)$, the powerset of n, by joining $X, Y \in \mathcal{P}(n)$ by an edge if and only if $X \cap Y = \emptyset$. We show that, for n large enough, $G = \mathcal{P}(n)$ satisfies $G \to (F)_r^e$.

For a family of disjoint subsets $S_1, \ldots, S_m \subseteq n$, where $m \geq v_F + \binom{v_F}{2}$, it is an easy exercise to construct a function $f : v_F \to \mathcal{P}(m)$ such that the subgraph of $\mathcal{P}(n)$ induced by vertices $\bigcup_{j \in f(0)} S_j, \ldots, \bigcup_{j \in f(v_F-1)} S_j$ is isomorphic to F. We call such a function a F-*embedding* function.

Note that an edge in $\mathcal{P}(n)$ is given by two mutually disjoint and nonempty subsets of n. Hence, edges correspond to 2-parameter words over the alphabet $A = \{0\}$ and vice versa (cf. Sect. 3.1.2). Applying the Graham-Rothschild theorem (Theorem 5.1) we deduce that for every r-coloring of the edges of $\mathcal{P}(n)$ there exists a family of disjoint subsets $S_1, \ldots, S_m \subseteq n$ such that every 2-parameter word given by these subsets has the same color. Let $X_i = \bigcup_{j \in f(i)} S_j$ for $0 \leq i < v_F$, where f is the F-embedding function. The sets X_i thus induce an subgraph $\hat{F}$ that is isomorphic to F. Every edge in this subgraph corresponds to a 2-parameter word given by some of the subsets S_i. Thus every edge in $\hat{F}$ has the same color. That is, $\hat{F}$ is the desired monochromatic induced F-subgraph.   □

Stronger versions of Theorem 12.4, along the lines of Folkman's Theorem 12.3, are also known. Namely, Nešetřil and Rödl (1976b) proved that there exists a graph G such that its clique number is equal to the clique number of F. In Chap. 16 we will prove this result for the special case when $F - K_\ell$. For now we continue with some negative results. We start with a simple lemma which shows that negative results for the vertex case imply negative results for the edge case as well.

Lemma 12.5. *Let F be a connected graph with chromatic number at least $r + 1$, for some $r \geq 2$. Then $G \nrightarrow (F)_r^v$ implies $G \nrightarrow (F)_2^e$.*

Proof. If $G \nrightarrow (F)_r^v$ then there exists an r-coloring of the vertices, that is, a partition $V = V_1 \cup \ldots \cup V_r$ such that no $G[V_i]$ contains a copy of F. We can thus color all edges in $G[V_i]$, for $0 \leq i < r$, with blue and all edges in $E(V_i, V_j)$, for $0 \leq i < j < r$, with red without inducing a monochromatic copy of F (as F is connected and is not r-partite).   □

Our next aim is to show that if G is too sparse then it cannot arrow F. To prepare for its proof we start with two beautiful partition results from graph theory. Recall that $G \nrightarrow (F)_r^e$ means that there exists an edge coloring with r-colors that avoids a monochromatic F in all colors. Nash-Williams' so-called Arboricity Theorem generalizes this question from a single graph to a graph class. Namely, he studies the question when can we find an r-coloring of the edges so that no color class contains *any* cycle. In fact, he determines the exact $r = r(F)$ so that this is possible.

Notation. The *arboricity* of a graph $G = (V, E)$ on at least two vertices is defined by

$$ar(G) = \max_{J \subseteq G, \, v_J > 1} \frac{e_J}{v_J - 1}.$$

One easily checks that for all graphs G with $m(G) \geq 1$ we have

$$m(G) \leq ar(G) \leq m(G) + \tfrac{1}{2}. \tag{12.1}$$

Theorem 12.6 (Nash-Williams' Arboricity Theorem). *For every graph $G = (V, E)$ there exists a partition of the edges into $\lceil ar(G) \rceil$ parts, $E = E_1 \cup \ldots \cup E_{\lceil ar(G) \rceil}$ such that all E_i are forests.*

Proof. We follow Chen et al. (1994). We call a partition of the edges $E_1 \cup \ldots \cup E_x = E$ such that all E_i are forests a *forest decomposition* of size x. A *minimal* forest decomposition is one for which x is as small as possible. Let $\rho(G)$ denote the number of sets in a minimal forest decomposition of G. We need to show that $\rho(G) = \lceil ar(G) \rceil$. One easily checks that we always have $\rho(G) \geq \lceil ar(G) \rceil$. So we only need to show that $\rho(G) \leq \lceil ar(G) \rceil$. Assume not. Then there exists a minimum counterexample G_0, minimum with respect to $v_{G_0} + e_{G_0}$, such that $\rho(G_0) > \lceil ar(G_0) \rceil$. We will show that this leads to a contradiction. For that we will show:

$(\star)$ For all $e \in E(G_0)$ the following holds. If $E_1, \ldots, E_{\rho(G_0)-1}$ is a forest decomposition of $G_0 - e$, then all E_i are spanning trees.

Before proving this let us see how this concludes the proof of the theorem. Clearly, $(\star)$ implies

$$e_{G_0} - 1 = (\rho(G_0) - 1) \cdot (v_{G_0} - 1).$$

Thus the assumption $\rho(G_0) > \lceil ar(G_0) \rceil$ implies

$$\rho(G_0) > \lceil ar(G_0) \rceil \geq \lceil \tfrac{e_{G_0}}{v_{G_0}-1} \rceil = \lceil (\rho(G_0) - 1) + \tfrac{1}{v_{G_0}-1} \rceil = \rho(G_0),$$

a contradiction. (In the last step we used that $\rho(G_0)$ is integral.)

So it remains to prove $(\star)$. To simplify notation we use in the following ρ to denote $\rho(G_0)$. If $(\star)$ is not true, then there exists $e \in E(G_0)$, $X \subsetneq V(G_0)$ and a forest decomposition $E_1, \ldots, E_{\rho-1}$ such that E_1 has a component with vertex set X, $E_1 + e$ contains a cycle C and this cycle is contained in $\hat{G} := G_0[X]$. As G_0 was a minimum counterexample we know that there exists a forest decomposition $A_1, \ldots, A_{\rho-1}$ of $\hat{G}$.

With these preparations at hand we proceed as follows. Consider all forest decompositions $\bar{E}_1, \ldots, \bar{E}_\rho$ of G_0 that have the following properties: (i) $\bar{E}_\rho = \{\bar{e}\}$ for some $\bar{e} \in E(\hat{G})$, and (ii) $\bar{E}_1$ has a component with vertex set exactly X. From all such decompositions (note that our original decomposition $E_1, \ldots, E_{\rho-1}$ satisfies both constraints, so this set is non-empty) we fix one that maximizes

$$\sum_{i=1}^{\rho-1} |A_i \cap \bar{E}_i|. \tag{12.2}$$

As $\bar{e} \in E(\hat{G})$, there exists $1 \leq t \leq \rho - 1$ such that $\bar{e} \in A_t$. Clearly, $\bar{e}$ induces a cycle C in $\bar{E}_t + \bar{e}$ (as otherwise we would have a forest decomposition of size $\rho - 1$ for G_0). We first argue that $C \subseteq E(\hat{G})$. Note that this is clearly the case if $t = 1$, by assumption (ii). So assume $t > 1$. As $\bar{e} \in E(\hat{G})$ by assumption (i), $C \nsubseteq E(\hat{G})$ implies that there exists $f \in C$ such that $|f \cap V(\hat{G})| = 1$. But then assumption (ii) implies that $E_1 + f$ is cycle-free. By construction $E_t + \bar{e} - f$ is also cycle-free, and we thus obtain a forest decomposition of size $\rho - 1$, which can't be. So we know that $C \subseteq E(\hat{G})$. As A_t is acyclic and contains $\bar{e}$, C has to contain an edge f that is not contained in A_t. Now consider $\bar{E}'_t := \bar{E}_t + \bar{e} - f$, $\bar{E}'_\rho := \{f\}$ and $\bar{E}'_i = \bar{E}_i$ for $i \neq t, \rho$. Then $\bar{E}'_1, \ldots, \bar{E}'_\rho$ is a forest decomposition that satisfies properties (i) and (ii) and for which the term in (12.2) is larger than for $\bar{E}_1, \ldots, \bar{E}_\rho$. This contradiction thus concludes the proof of the theorem. $\square$

The next theorem generalizes Nash-Williams' Arboricity Theorem from forests to components that contain at most one cycle.

Theorem 12.7. *For every graph $G = (V, E)$ there exists a partition of the edges into $\lceil m(G) \rceil$ parts, $E = E_1 \cup \ldots \cup E_{\lceil m(G) \rceil}$ such that for all E_i the following holds: all components of E_i contain at most one cycle.*

Proof. Construct a bipartite graph $\hat{G} = (A \cup B, \hat{E})$ as follows. The set A consists of $\lceil m(G) \rceil$ copies of each vertex in G. The set B consists of all edges in G and we connect each element in B to all copies of the two vertices in this edge. We claim that $\hat{G}$ contains a matching M that covers all vertices in B. To see this it suffices to check Hall's condition. Let $E' \subseteq E(G)$ be arbitrary and let $v_{E'} := |V(E')|$ denote the number of vertices that are contained in the edges from E'. Then

$$|\Gamma_{\hat{G}}(E')| = \lceil m(G) \rceil \cdot v_{E'} \geq \frac{\lceil m(G) \rceil}{m(G)} \cdot |E'| \geq |E'|,$$

where the inequality $|E'|/v_{E'} \leq m(G)$ follows from the definition of the m-density. Thus, the desired matching exists.

Note that a matching that covers all vertices in B (and thus all edges in G) defines a partition of $E(G)$ into $\lceil m(G) \rceil$ subgraphs constructed as follows. For the ith subgraph consider the ith copy of each vertex of G. If the copy is matched put the corresponding edge into the ith subgraph. Note that, by construction, these subgraphs have the following property: for every $X \subseteq V(G)$ the graph induced by X contains at most $|X|$ edges; thus all these subgraphs have m-density at most 1, and their components can thus contain at most one cycle. $\square$

These results easily imply some negative Ramsey results:

Lemma 12.8. *Let F and G be two graphs such that there exists an $\epsilon > 0$ so that $m(G) \leq r \cdot \lfloor m(F) - \epsilon \rfloor$ or $ar(G) \leq r \cdot \lfloor ar(F) - \epsilon \rfloor$. Then $G \nrightarrow (F)^e_r$.*

Proof. Assume first that $m(G) \leq r \cdot \lfloor m(F) - \epsilon \rfloor$. As $r \cdot \lfloor m(F) - \epsilon \rfloor$ is integral, we can replace the assumption of the lemma with the stronger statement

$r \cdot \lfloor m(F) - \epsilon \rfloor \geq \lceil m(G) \rceil$. Theorem 12.7 thus implies that we can partition the edges of G into $r \cdot \lfloor m(F) - \epsilon \rfloor$ subgraphs in such a way that all these subgraphs have m-density at most 1. We color the first $\lfloor m(F) - \epsilon \rfloor$ of these subgraphs with color 0, the next $\lfloor m(F) - \epsilon \rfloor$ with color 1 and so on. Then for all i the subgraphs with color i have m-density at most $\lfloor m(F) - \epsilon \rfloor < m(F)$ and can thus not contain a copy of F.

The case $ar(G) \leq r \cdot \lfloor ar(F) - \epsilon \rfloor$ follows analogously, using Theorem 12.6 instead of Theorem 12.7.   □

We conclude this section with the observation that in order to arrow F the graph G must have a suitable minimum degree. Recall that $\delta(G)$ denotes the minimum degree in G, i.e., $\delta(G) := \min_{v \in V(G)} \deg(v)$ and $\delta_{max}(G) := \max_{G' \subseteq G} \delta(G')$ denotes the maximum minimum degree in all subgraphs of G.

Lemma 12.9. *Let F and G be two graphs such that $\delta_{max}(G) \leq r(\delta(F) - 1)$. Then $G \nrightarrow (F)_r^e$.*

Proof. Construct a sequence $v_1, v_2, \ldots, v_{v_G}$ as follows: let v_i be a vertex of minimum degree in $G - \{v_1, \ldots, v_{i-1}\}$. Then every vertex v_i has degree at most $\delta_{max}(G)$ into $G[\{v_{i+1}, \ldots, v_{v_G}\}]$, the graph induced by the vertices 'to the right'. Now color the vertices of G 'backwards', i.e., starting with v_{v_G} (which is colored arbitrarily). As every vertex v_i has degree at most $\delta_{max}(G) \leq r(\delta(F) - 1)$ into the part that is already colored, we can partition the edges incident to v_i that end in the part that is already colored into r groups, each with at most $\delta(F) - 1$ edges, and color the ith group with the color i. Clearly, the colored part can then not contain a monochromatic copy of F that contains v_i. By repeating this procedure for every vertex v_i we thus obtain a coloring of G without a monochromatic F-subgraph.   □

12.3 Colorings of Subgraphs

The results mentioned so far are in a sense just variations of the pigeonhole principle, but they have the typical flavor of what might be expected for graphs in general. First we try to establish an induced graph partition theorem and then we exploit all possible restrictions. A Ramsey theorem for partitioning complete subgraphs was obtained by Deuber (1975) and by Nešetřil and Rödl (1975).

Theorem 12.10. *Let F be a finite graph and let ℓ and r be positive integers. Then there exists a finite graph G satisfying $G \xrightarrow{ind} (F)_r^{K_\ell}$.*

Proof. This theorem follows easily by adapting the proof of Theorem 12.4. Recall that there we defined a graph on the powerset $\mathcal{P}(n)$ of n by joining two sets $X, Y \in \mathcal{P}(n)$ by an edge if and only if $X \cap Y = \emptyset$. Note that the graph defined on $\mathcal{P}(n)$ has the property than every ℓ pairwise disjoint sets form a complete subgraph. Hence, K_ℓ-subgraphs correspond to ℓ-parameter words over the alphabet $A = \{0\}$ and vice versa. We can thus complete the proof analogously as in the proof of

Theorem 12.4 by applying the Graham-Rothschild Theorem 5.1, this time for ℓ instead of 2.   □

By reversing the notion of edge and non-edge (i.e., considering the complementary graph) Theorem 12.10 immediately implies an analogous result with respect to partitions of stable subsets on k vertices (so called empty graphs, where empty means that we do not have any edges).

What about graphs which are neither complete nor empty? The interesting fact is that there is no Ramsey type theorem for such graphs.

We illustrate this at the smallest connected graph which is neither complete nor empty, viz., P_3, a path on three vertices. Let us denote by C_4 a cycle of length 4.

Proposition 12.11. *For every graph G there exists a 2-coloring of the P_3-subgraphs such that no C_4-subgraph is monochromatic.*

Proof. Given $G = (V, E)$, impose an arbitrary (but fixed) total order on the set of vertices and color a P_3-subgraph P' with color 0 if its degree two vertex (i.e., the vertex in the middle) is the largest vertex of P', largest with respect to the total order, and color P' with color 1 otherwise. As every C_4-subgraph of G contains P_3-subgraphs of both kinds, we see that no monochromatic C_4-subgraph exists.   □

The idea in this argument can be extended to prove the following more general result, due to Nešetřil and Rödl (1975).

Theorem 12.12. *Let H be a finite graph which is neither complete nor empty. Then there exists a finite graph F such that $G \nrightarrow (F)_2^H$ for every finite graph G.*

It seems natural to try to extend the idea of using orderings from the proof of Proposition 12.11 to also obtain a proof of Theorem 12.12. As it turns out this works in principle, but the details are technically quite involved. Before proving the negative result from Theorem 12.12 we prove a positive result about another structure, namely *ordered graphs*. In a second step we then show how such a positive result can be used in order to prove Theorem 12.12.

At this point we introduce the concept of ordered graphs. An ordered graph is a triple $(V, E, \leq)$, where (V, E) is a graph and $\leq$ is a total order on the set V of vertices. An embedding between ordered graphs $(V, E, \leq)$ and $(V', E', \leq')$ is given by a monotone injection $f : V \to V'$, i.e., $x < y$ implies $f(x) < f(y)$, which is also a graph isomorphism.

Observe that contrary to ordinary graphs ordered graphs are rigid. In particular, nontrivial automorphisms do not exist. Moreover, the kind of counterexample showing that P_3 does not have the Ramsey property is not available any more. In fact, this rigidity property of ordered graphs allows us to obtain Ramsey statements also in cases where the corresponding result for unordered graphs does not exist.

More generally, one can show that the class of finite ordered graphs is a Ramsey class, i.e., it admits a complete analogue of Ramsey's theorem. This has been established independently by Abramson and Harrington (1978) and by Nešetřil and

Rödl (1977, 1983b). A proof using Graham-Rothschild parameter sets has been given by Prömel and Voigt (1981a).

Theorem 12.13 (Ramsey theorem for ordered graphs). *Let $(H, \leq)$ and $(F, \leq)$ be finite ordered graphs and let r be a positive integer. Then there exists a finite ordered graph $(G, \leq)$ such that $(G, \leq) \overset{ind}{\to} (F, \leq)_r^{(H, \leq)}$, i.e., for every r-coloring of the ordered $(H, \leq)$-subgraphs of $(G, \leq)$ there exists a monochromatic ordered $(F, \leq)$-subgraph.*

Proof. We use a similar construction as the one given in the proof of Theorem 12.4, only this time joining two subsets by an edge if and only if they intersect. That is, let $G = (\mathcal{P}(n), E)$ be the graph defined on the powerset of n by joining $X, Y \in \mathcal{P}(n)$ by an edge if and only if $X \cap Y \neq \emptyset$. We now define an ordering of the vertices of G. For two elements $X, Y \in \mathcal{P}(n)$ that are not contained in each other (that is, neither $X \subseteq Y$ nor $Y \subseteq X$ holds), we let $X \leq Y$ if and only if $\min X \setminus Y < \min Y \setminus X$. One easily checks that this defines a partial order on $\mathcal{P}(n)$. We extend this partial order arbitrarily to obtain a linear order on $\mathcal{P}(n)$. With slight abuse of notation we let $\leq$ denote this linear order. We show that the ordered graph $(G, \leq)$, for n large enough, satisfies $(G, \leq) \overset{ind}{\to} (F, \leq)_r^{(H, \leq)}$.

Let $(h_0, \ldots, h_{e_H-1})$, resp. $(g_0, \ldots, g_{e_F-1})$, be the lexicographic enumeration of the edges of H and F. For ease of notation, in the remainder of the proof we will omit the explicit statement of the fact that all graphs considered in this proof are ordered. That is, we will simply write H instead of $(H, \leq)$ and speak of colorings of H-subgraphs when we actually mean colorings of $(H, \leq)$-subgraphs, etc.

Let $\Delta : \binom{G}{H} \to r$ be a given r-coloring of the H-subgraphs of G. We define an r-coloring $\Delta^* : [1]\binom{n}{v_H+e_H} \to r$ of $(v_H + e_H)$-parameter words over the alphabet $1 = \{0\}$ as follows. Given an $(v_H + e_H)$-parameter word $a \in [1]\binom{n}{v_H+e_H}$, let $S_i = \{j \mid a_j = \lambda_i\}$ for $0 \leq i < v_H + e_H$. Clearly, all sets S_i are pairwise disjoint, and since $\min \lambda_i^{-1} < \min \lambda_j^{-1}$ whenever $i < j$, we also have $S_i \leq S_j$ (interpreting the sets S_i as vertices of G). Furthermore, let

$$X_i = S_i \cup \bigcup_{j \,:\, i \in h_j} S_{v_H+j}$$

for $0 \leq i < v_H$. It is easy to see that, by construction, $X_0, \ldots, X_{v_H-1}$ induce an H-subgraph H^* of G. We use this H-subgraph to define the coloring for a, i.e., we let $\Delta^*(a) := \Delta(H^*)$.

For all $n \geq GR(v_F + e_F, v_H + e_H, r)$ we can apply the Graham-Rothschild theorem (Theorem 5.1) to the coloring Δ^* to deduce that there exists a $(v_F + e_F)$-parameter word $f \in [1]\binom{n}{v_F+e_F}$ such that

$$\Delta^*(f \cdot g) = \Delta^*(f \cdot g') \tag{12.3}$$

for every $g, g' \in [1]\binom{v_F+e_F}{v_H+e_H}$. Similarly as before, let $S_i = \{j \mid f_j = \lambda_i\}$ for $0 \leq i < v_F + e_F$ and

$$F_i = S_i \cup \bigcup_{j \,:\, i \in g_j} S_{v_F + j}.$$

It is easy to see that $F_0, \ldots, F_{v_F-1}$ induce an F-subgraph F^* in G. We claim that F^* is monochromatic with respect to its H-subgraphs.

Let H^* be any H-subgraph of F^*, and let $(X_0, \ldots, X_{v_H-1})$ be an enumeration of the vertices of H^*, monotone with respect to $\leq$. For $0 \leq j < e_H$ set $S'_{v_H+j} :=$ $X_a \cap X_b$, where $\{a, b\} = h_j$, and $S'_i := X_i \setminus \bigcup_{j \,:\, i \in h_j} S'_{v_H+j}$ for $0 \leq i < v_H$. Since edges of both F and H are enumerated lexicographically and the vertices of H^* are a subset of those of F^* (and both sets are ordered increasingly), it easily follows that $S'_i < S'_j$ whenever $i < j < v_H + e_H$. Consider now $g \in [1]\binom{v_F+e_F}{v_H+e_H}$ defined by

$$g_j = \begin{cases} \lambda_i, & j \in S'_i \text{ for some } 0 \leq i < v_H + e_H, \\ 0, & \text{otherwise,} \end{cases}$$

Let H^{**} be an H-subgraph constructed from $f \cdot g$, as before. It is easy to see that $H^* = H^{**}$, thus by the definition of coloring Δ^* we have $\Delta^*(f \cdot g) = \Delta(H^{**}) = \Delta(H^*)$. Therefore, by (12.3) all H-subgraphs of F^* are monochromatic, which finishes the proof. □

This brings us also to another concept, the *ordering property*. The ordering property allows to deduce Ramsey type results for unordered graphs. This concept has been introduced by Nešetřil and Rödl (1975) and was studied further in Nešetřil and Rödl (1978a).

Theorem 12.14 (Ordering property of finite graphs). *Let $(H, \leq)$ be a finite ordered graph. Then there exists a finite graph G (unordered yet) such that for every total order $\leq^*$ of the vertices of G there exists an ordered embedding of $(H, \leq)$ into $(G, \leq^*)$. (To have a shorthand abbreviation we write $G \xrightarrow{ord} (H, \leq)$ for this property of G).*

Proof. Let $\{v_0, \ldots, v_{m-1}\}$ be an enumeration of the vertices of H, monotone with the respect to the order $\leq$. Without loss of generality we may assume that all pairs $\{v_{i-1}, v_i\}$ are edges of H (otherwise, this can be easily achieved by adding vertices of degree two). Let $(H, \leq^{-1})$ be a copy of H with the order $\leq^{-1}$ being reverse to $\leq$. Let the ordered graph $(H^*, \leq^*)$ be the disjoint union of $(H, \leq)$ and $(H, \leq^{-1})$ (where, say, all vertices of $(H, \leq)$ precede the vertices from $(H, \leq^{-1})$). Let $(G, \leq')$ be a finite ordered graph satisfying $(G, \leq') \rightarrow (H^*, \leq^*)_2^e$, which exists by Theorem 12.13.

We claim that the graph G, i.e., the unordered version of $(G, \leq')$, satisfies the requirements of Theorem 12.14. Let $\leq''$ be an arbitrary total order of the vertices of G. We define a 2-coloring on the edges of G as follows: color an edge $\{x, y\}$ with color 0 if $x \leq' y$ and $x \leq'' y$ (the two orders agree) and color it with color 1 otherwise. By choice of $(G, \leq')$ there exists an ordered $(H^*, \leq^*)$-subgraph with all its edges in the same color. If this color is 0 the $(H, \leq)$ part yields an order preserving $(H, \leq)$-subgraph of $(G, \leq)$ and if this color is 1 the $(H, \leq^{-1})$ part of $(H^*, \leq^*)$ yields an order preserving subgraph of $(G, \leq'')$.   □

Next we show how the ordering property of finite graphs can be applied in proving Theorem 12.12.

Proof of Theorem 12.12. First observe that if H is a finite graph which is neither complete nor empty then there exist two total orders $\leq_1$ and $\leq_2$ on the vertices of H such that there is no order preserving graph embedding between $(H, \leq_1)$ and $(H, \leq_2)$. Applying Theorem 12.14 on a graph obtained from the disjoint union of $(H, \leq_1)$ and $(H, \leq_2)$ (where, as in the previous proof, all vertices of $(H, \leq_1)$ precede the vertices from $(H, \leq_2)$), we get a finite graph F be such that $F \overset{\text{ord}}{\to} (H, \leq_1)$ and $F \overset{\text{ord}}{\to} (H, \leq_2)$.

We claim that $G \not\to (F)_2^H$ for every graph G, i.e., for every graph G there exists a 2-coloring of its H-subgraphs such that no F-subgraph is monochromatic. For defining such a 2-coloring we first impose an arbitrary (but fixed) total order $\leq$ on the vertices of F. Now we color all H-subgraphs of G which are embeddings of $(H, \leq_1)$ into $(G, \leq)$ with color 0 and all other H-subgraphs with color 1. By choice of F, every F-subgraph of G contains H-subgraphs with order pattern $\leq_1$ as well as H-subgraphs with order pattern $\leq_2$. So no monochromatic F-subgraph exists.

□

Let us summarize the results of Theorems 12.10 and 12.12.

Notation. Let H be a finite (unordered) graph. We say that H has the *partition property* with respect to the class of finite (unordered) graphs if for every finite graph F and every positive integer r there exists a finite graph G satisfying $G \overset{\text{ind}}{\to} (F)_r^H$.

Corollary 12.15. *Let H be a finite graph. The following assertions are equivalent:*

(1) *H has the partition property with respect to the class of finite graphs,*
(2) *H is complete or empty,*
(3) *There exists a total order $\leq$ on the vertices of H s.t. $H \overset{\text{ord}}{\to} (H, \leq)$.*   □

12.4 Unbounded Colorings of Subgraphs

As most of the graphs occurring in this section are ordered graphs we simple say that $(G, \leq)$, resp., $(F, \leq)$ is an ordered graph, without explicitly distinguishing the tacit underlying orders of the vertices of G, resp., F. If it is clear from the context that we deal with ordered graphs we also write simply G, resp., F.

The result of this section is a canonizing version of the Ramsey theorem for ordered graphs (Theorem 12.13). As it turns out, this canonizing version is quite similar to the Erdős-Rado canonization theorem.

Notation. Let $H = (V, E)$ be an ordered graph and let $J \subseteq v_G$ be a (possibly empty) set. By $H : J$ we denote the subgraph of H induced by the vertices $V : J$. Recall that V itself is an ordered set and, hence, the notion of its J-subset is well-defined.

Theorem 12.16. *Let $(H, \leq)$ and $(F, \leq)$ be finite ordered graphs. Then there exists a finite ordered graph $(G, \leq)$ such that*

$$(G, \leq) \xrightarrow{can} (F, \leq)^{(H, \leq)},$$

i.e., for every (unbounded) coloring of the ordered $(H, \leq)$-subgraphs of $(G, \leq)$ there exists an ordered $(F, \leq)$-subgraph $\tilde{F}$ of $(G, \leq)$ and there exists a set $J \subseteq k$, where k is the number of vertices of H, such that

$$\Delta(H') = \Delta(H'') \quad \text{if and only if} \quad H' : J = H'' : J$$

holds for all ordered $(H, \leq)$-subgraphs H' and H'' of $\tilde{F}$.

This theorem is due to Prömel and Voigt (1985). Independently, it has also been observed by Nešetřil and Rödl (unpublished). In this section we show how Theorem 12.16, the canonizing Ramsey theorem for ordered graphs, can be deduced from the Ramsey theorem for ordered graphs (Theorem 12.13). The proof goes by induction on the number of vertices of H and the idea is very similar to the idea used in proving the Erdős-Rado canonizing theorem.

Proof of Theorem 12.16. We use the following notation. Let H be an ordered graph on k vertices, and assume $x_0 < x_1 < \dots < x_{k-1}$ are the vertices given in the increasing order. For $i < k$ we denote by $H + i$ the graph obtained from H by doubling vertex x_i, i.e., we add a new vertex y_i with $x_i < y_i < x_{i+1}$ and for each edge $\{x_i, x_j\}$ of H we add the edge $\{y_i, x_j\}$. Finally, we denote with $F' \cup F''$ the ordered graph which is a disjoint union of ordered graphs $(F', \leq')$ and $(F'', \leq'')$, with vertices of F' preceding vertices of F''.

Clearly, $(H + i) : ((k + 1)\backslash\{i + 1\})$ as well as $(H + i) : ((k + 1)\backslash\{i\})$ are isomorphic to H, so the additional vertex y_i plays the same rôle as x_i.

Now we prove Theorem 12.16 by induction on the number of vertices of H. If H is a single vertex the result follows from a result of Nešetřil and Rödl (Theorem 16.8) that we will prove in Chap. 16.

Let $(H, \leq)$ be a graph on k vertices and assume by induction that Theorem 12.16 is valid for all ordered graphs H' on $k - 1$ vertices.

Let $(F, \leq)$ be an ordered graph. We construct an ordered graph $(G, \leq)$ satisfying $(G, \leq) \xrightarrow{can} (F, \leq)^{(H, \leq)}$ as follows:

(1) Let $H_0, \dots, H_{s-1}$ be an enumeration of the ordered $(H, \leq)$-subgraphs of $(F, \leq)$. For every $\nu < s$ we enlarge F adding vertices y_i^ν for $i < k$ and adding

edges such that the subgraph spanned by the vertices of H_ν together with the vertex y_i^ν is isomorphic to $H + i$.

We call the resulting graph $\tilde{F}$. Let the index set $\tilde{I}$ be such that $\tilde{F} : \tilde{I} = F$.

(2) For every pair $\mu < \nu < s$ let $A(\mu, \nu)$ be the subgraph of F which is spanned by the union of the vertices of the H-subgraphs H_μ and H_ν. Let the index sets $I(\mu, \nu)$, resp., $J(\mu, \nu)$ be such that $A(\mu, \nu) : I(\mu, \nu) = H_\mu$, resp., $A(\mu, \nu) : J(\mu, \nu) = H_\nu$.

(3) According to repeated applications of the Ramsey theorem for graphs (Theorem 12.13) let the ordered graph B be such that for every family $\Delta_{\mu,\nu}$, $\mu < \nu < s$, of 2-colorings of the $A(\mu, \nu)$-subgraphs of B there exists a $\tilde{F}$-subgraph such that all colorings $\Delta_{\mu,\nu}$ restricted to the $A(\mu, \nu)$-subgraphs of this $\tilde{F}$-subgraph are monochromatic colorings.

(4) Let the ordered graph D be such that $D \stackrel{\text{can}}{\to} (F \cup B)^{H:(k\backslash\{i\})}$ for every $0 \le i \le k$. Such a graph D exists according to the inductive assumption.

(5) Again according to repeated applications of the Ramsey theorem for ordered graphs let the ordered graph G be such that for every family $\Delta_i, i < k$, of 2-colorings of the $(H + i)$-subgraphs of G there exists a D-subgraph such that all Δ_i restricted to the $(H + i)$-subgraphs of this D-subgraph are monochromatic colorings.

Now we claim that this graph G has the desired properties, i.e., it satisfies $G \stackrel{\text{can}}{\to} (F)^H$.

To verify this let Δ be an unbounded coloring of the ordered H-subgraphs of G. For every $i < k$ we define a 2-coloring Δ_i of the $(H + i)$-subgraphs of G by

$$\Delta_i(E) = \begin{cases} 0 & \text{if } \Delta(E : ((k + 1)\backslash\{i\})) = \Delta(E : ((k + 1)\backslash\{i + 1\})) \\ 1 & \text{otherwise.} \end{cases}$$

for every $(H + i)$-subgraph E.

By choice of G there exists a D-subgraph D^* such that $\Delta_i(E^*) = \Delta_i(E^{**})$ for every $i < k$ and all $(H + i)$-subgraphs E^* and E^{**} of D^*. We distinguish two cases, which will be handled separately:

Case (i):   For some $i < k$ the coloring Δ_i restricted to the $(H + i)$-subgraphs of D^* is monochromatic in color 0.

Case (ii):   All restrictions of the colorings Δ_i to $(H + i)$-subgraphs of D^* are monochromatic in color 1.

First assume that case (i) occurs. Let $i < k$ be such that

$$\Delta(E^* : ((k + 1)\backslash\{i\})) = \Delta(E^{**} : ((k + 1)\backslash\{i + 1\}))$$

for $(H + i)$-subgraphs E^* and E^{**} of D^*. (In other words, the Δ-color of H-subgraphs does not depend on the i-th vertex.) But this means that the (unbounded) coloring Δ^* which is defined on the $H : (k\backslash\{i\})$-subgraphs of D^*

by $\Delta^*(H^* \; : \; (k\backslash\{i\})) = \Delta(H^*)$ for every H-subgraph H^* of D^* is well-defined (those $H \; : \; (k\{i\})$-subgraphs which are not subgraphs of H in D^* we color arbitrarily). By choice of D there exists a subset $\tilde{J} \subseteq k - 1$ and there exists a F-subgraph F^* of D^* such that

$$\Delta^*(H^* : (k\backslash\{i\})) = \Delta^*(H^{**} : (k\backslash\{i\}))$$

$$\text{if and only if} \quad (H^* : (k\backslash\{i\})) : \tilde{J} = (H^{**} : (k\backslash\{i\})) : \tilde{J}$$

holds for all H-subgraphs H^* and H^{**} of F^*.

Let $J \subseteq k$ be such that $(H \; : \; (k\backslash\{i\})) \; : \; \tilde{J} = H \; : \; J$. By definition of Δ^* it follows for all H-subgraphs H^* and H^{**} of G^* that

$$\Delta(H^*) = \Delta(H^{**}) \qquad \text{if and only if}$$

$$\Delta^*(H^* : (k\backslash\{i\})) = \Delta^*(H^{**} : (k\backslash\{i\})) \qquad \text{if and only if}$$

$$(H^* : (k\backslash\{i\})) : \tilde{J} = (H^{**} : (k\backslash\{i\})) : \tilde{J} \qquad \text{if and only if}$$

$$H^* : J = H^{**} : J.$$

In other words, F^* and J have the desired properties.

Next assume that case (ii) occurs. Then we know that

$$\Delta(E^* : ((k + 1)\backslash\{i\})) \neq \Delta(E^{**} : ((k + 1)\backslash\{i + 1\})) \tag{12.4}$$

for all $i < k$ and all $(H + i)$-subgraphs E^* and E^{**} of D^*. (In other words, the Δ-color of H-subgraphs depends on the i-th vertex for each i).

Under these circumstances we show that there exists a F-subgraph on which Δ acts as a one-to-one coloring.

For every $\mu < \nu < s$ consider the 2-coloring $\Delta_{\mu,\nu}$ of the $A(\mu, \nu)$-subgraphs of D^* which is defined by

$$\Delta_{\mu,\nu}(A) = \begin{cases} 0 & \text{if } \Delta(A : I(\mu, \nu)) = \Delta(A : J(\mu, \nu)) \\ 1 & \text{otherwise.} \end{cases}$$

Since $D \xrightarrow{\text{can}} (F \cup B)^{H:(k\backslash\{i\})}$, we have that D^* contains a B-subgraph B^*. Further, by choice of B there exists a $\tilde{F}$-subgraph $\tilde{F}^*$ of B^* such that each $\Delta_{\mu,\nu}$ restricted to the $A(\mu, \nu)$-subgraphs of $\tilde{F}^*$ is a monochromatic coloring. We claim that each such restriction is monochromatic in color 1. We argue by contradiction. Assume to the contrary that some $\Delta_{\mu,\nu}$ is monochromatic in color 0. Let A be any $A(\mu, \nu)$-subgraph of $\tilde{F}^* : \tilde{I}$. Then

$$\Delta(A : I(\mu, \nu)) = \Delta(A : J(\mu, \nu)).$$

But, as $\mu \neq \nu$ we have that

$$A : I(\mu, \nu) \neq A : J(\mu, \nu).$$

Hence, some vertex of $A : J(\mu, \nu)$ does not belong to $A : I(\mu, \nu)$. Say, the i-th vertex of $A : J(\mu, \nu)$ has this property. By construction of $\tilde{F}$(cf.(1)), there exists a vertex y in $\tilde{F}^*$ which is not a vertex of $\tilde{F}^* : \tilde{I}$ such that the subgraph E induced by y plus the vertices of $A : J(\mu, \nu)$ is an $(H + i)$-subgraph of $\tilde{F}^*$, with $E : ((k + 1)\backslash\{i + 1\}) = A : J(\mu, \nu)$, and additionally, replacing in A the i-th vertex of $A : J(\mu, \nu)$ by y again yields an $A(\mu, \nu)$-subgraph. We call this later $A(\mu, \nu)$-subgraph A^*. Then $A^* : I(\mu, \nu) = A : I(\mu, \nu)$ and $A^* : J(\mu, \nu) = E : ((k + 1)\backslash\{i\})$.

According to our assumption on $\Delta_{\mu, \nu}$ it follows that

$$\Delta(A : I(\mu, \nu)) = \Delta(A : J(\mu, \nu))$$

and thus by choice of $\tilde{F}^*$ also

$$\Delta(A^* : I(\mu, \nu)) = \Delta(A^* : J(\mu, \nu)).$$

Hence $\Delta(E : (k + 1)\backslash\{i\})) = \Delta(E : ((k + 1)\backslash\{i + 1\}))$, but this contradicts the general assumption of (12.4).

So we know that every coloring $\Delta_{\mu, \nu}$ restricted to the $A(\mu, \nu)$-subgraphs of $\tilde{F}^*$ is monochromatic in color 1. But this means that Δ restricted to the H-subgraphs of $F^* = \tilde{F}^* : \tilde{I}$ is one-to-one. In other words F^* and $J = k$ have the desired properties. $\square$

Chapter 13
Infinite Graphs

Considering infinite graphs, the picture is, even in the case of countable graphs, far from being complete. We discuss some of the pieces which are known. Section 13.1 deals with vertex colorings of Rado's graph R. We show that $R \rightarrow (R)_r^v$ for every positive integer r. In Sect. 13.2 we consider K_ℓ-free subgraphs of Rado's graph. Section 13.3 is concerned with edge colorings. Most of the results of this section are contained in the important paper (Erdős et al. 1975). We show that countable graphs do not have the edge partition property.

Graphs and embeddings of graphs are defined as in Chap. 12. The cardinality of a graph is the cardinality of its vertex set. The Ramsey arrow is used as introduced in Chap. 12.

13.1 Rado's Graph

Rado (1964) describes a construction of a universal countable graph, let us call it R, which has a lot of interesting properties. Being universal means that every countable graph can be embedded into R. The crucial property of R is that it is ω-good.

Definition 13.1. A graph $G = (V, E)$ is ω-good if for any two finite and disjoint sets X and Y of vertices there exists a vertex z not belonging to $X \cup Y$ such that z is joined by an edge to all $x \in X$ and not joined to any $y \in Y$.

Proposition 13.2. *Let $G = (V, E)$ be an ω-good graph. Then G is universal for countable graphs, i.e., every countable graph can be embedded into G.*

Proof. Let F be a countable graph. Without loss of generality we assume that $F = (\omega, E_F)$, i.e., the vertices of F are the nonnegative integers. We construct an embedding $f : \omega \rightarrow V$ inductively, one vertex at a time.

Let $f(0)$ be any vertex in V and suppose that $f(0), \ldots, f(n-1)$ have been defined. Consider the vertex n and let $A = \{k < n \mid \{k, n\} \in E_F\}$ be the set of

H.J. Prömel, *Ramsey Theory for Discrete Structures*,
DOI 10.1007/978-3-319-01315-2_13,
© Springer International Publishing Switzerland 2013

previous vertices which are joined to n, resp., let $B = \{k < n \mid \{k, n\} \notin E_F\}$ be its complement. Let $X = f(A)$ and $Y = f(B)$ be the corresponding sets in G. As G is ω-good there exists a vertex $z \in V \setminus (X \cup Y)$ which is joined to all $x \in X$ and not joined to any $y \in Y$. So, define $f(n) = z$ for any such z and continue as before. □

Actually, the above proof establishes slightly more, namely: any embedding of a finite subgraph of F into G can be extended to an embedding of F into G. This is to say that the automorphism group of any ω-good graph acts transitively on finite subgraphs, this property is sometimes called ultrahomogeneity.

Using the argument in the proof of Proposition 13.2 *back and forth* yields:

Proposition 13.3. *Any two countable ω-good graphs F and G are isomorphic.*

Proof. Proceed as in the proof of Proposition 13.2, however, 'back and forth'. At even-numbered steps try to embed F into G and at odd-numbered steps try to embed G into F. Eventually, any f constructed in such a way is an isomorphism.

□

Knowing that, up to isomorphisms, there is just one countable ω-good graph we call this graph Rado's graph R. Still we are lacking some kind of explicit description, resp. a proof of the existence of countable ω-good graphs. Such an explicit construction has been given in Rado (1964).

Definition 13.4. Let the set $\mathcal{R} \subseteq [\omega]^2$ be defined as follows. Given $k < m$ put $\{k, m\} \in \mathcal{R}$ if and only if 2^k occurs in the binary expansion of m. Let $R = (\omega, \mathcal{R})$ be the graph which has as vertices nonnegative integers and $\mathcal{R}$ as the set of edges. One easily observes that this graph R is, in fact, ω-good.

Remark 13.5. About at the same time when Rado gave his construction, Erdős and Rényi (1963) showed that if one considers countably infinite random graphs by inserting edges independently with probability $1/2$ then almost surely any such random graph is ω-good. Thus, almost surely a countable random graph is isomorphic to Rado's graph R. For further interesting properties of Rado's graph compare, e.g., Cameron (1984).

Theorem 13.6. *For every positive integer r we have*

$$R \overset{ind}{\to} (R)^v_r.$$

Proof. Let $\Delta : \omega \to r$ be a coloring of the vertices of R. Let $V_i := \{n < \omega \mid \Delta(n) = i\}$ denote the set of vertices that are colored with color i. If the graph induced by V_i is ω-good then Proposition 13.2 implies that it contains an induced R-subgraph which is monochromatic in color i and we are done. Otherwise there exist finite and disjoint sets $X_i, Y_i \subseteq V_i$ such that V_i contains no vertex that is connected to all vertices in X_i and to no vertex in Y_i. Now assume that no set V_i induces an ω-good graph. Then consider $X := \bigcup_{i<r} X_i$ and $Y := \bigcup_{i<r} Y_i$. By construction, X and Y are finite and disjoint. As the Rado graph is ω-good, there exists a vertex z that is connected to all vertices in X and to no vertex in Y. As z has to be colored with some color this contradicts the definition of the sets X_i and Y_i. $\qquad\square$

With slightly more effort we also obtain a canonical version:

Theorem 13.7. *For every (unbounded) coloring $\Delta : \omega \to \omega$ of the vertices of the Rado graph $R = (\omega, \mathcal{R})$, there exists $X \subseteq \omega$ spanning a subgraph which is isomorphic to Rado's graph such that $\Delta \!\upharpoonright\! X$ is constant or one-to-one.*

Proof. For finite and disjoint sets X and Y in ω, let $\Gamma(X, Y)$ be the set of vertices that are joined by an edge to all vertices in X and to no vertex in Y,

$$\Gamma(X, Y) = \{z < \omega \mid z \notin X \cup Y, \{x, z\} \in \mathcal{R} \text{ for all } x \in X,$$

$$\{y, z\} \notin \mathcal{R} \text{ for all } y \in Y\}.$$

We first prove that for any finite and disjoint sets X and Y in ω, $\Gamma(X, Y)$ spans a ω-good graph. Assuming otherwise, there exists finite (and disjoint) subsets $C_1, C_2 \subseteq \Gamma(X, Y)$, for some finite (and disjoint) $X, Y \subseteq \omega$, such that $\Gamma(C_1, C_2) \cap \Gamma(X, Y) = \emptyset$. Since R is ω-good there exists some z not in $X \cup C_1 \cup Y \cup C_2$ such that z is joined to all vertices in $X \cup C_1$ and to no vertex in $Y \cup C_2$. But then $z \in \Gamma(X \cup C_1, Y \cup C_2) = \Gamma(C_1, C_2) \cap \Gamma(X, Y)$, yielding the desired contradiction.

Using this observation, we inductively find a set of vertices $\{x_0, \ldots, x_{n-1}\}$ such that

(1) $\Delta(x_i) \neq \Delta(x_j)$ for all $i < j < n$,
(2) $\{x_0, \ldots, x_{n-1}\}$ spans a graph which is isomorphic to the one spanned by $\{0, \ldots, n-1\}$; in other words, $\{x_0, \ldots, x_{n-1}\}$ yields a one-to-one colored initial segment of R,

or deduce that there exists a monochromatic subgraph isomorphic to Rado's graph. Note that for $n = 0$ these assertions hold vacuously, yielding the beginning of the induction. Having vertices $\{x_0, \ldots, x_{n-1}\}$ which satisfy (1) and (2), let $A_n \subseteq \{x_0, \ldots, x_{n-1}\}$ resp., $B_n = \{x_0, \ldots, x_{n-1}\} \setminus A_n$ be such that for every $x \in \Gamma(A_n, B_n)$ the set $\{x_0, \ldots, x_{n-1}, x\}$ is isomorphic to $\{0, \ldots, n\}$. If there exists a vertex $x \in \Gamma(A_n, B_n)$ such that $\Delta(x) \neq \Delta(x_i)$ for $0 \leq i < n$, then setting $x_n = x$ finishes the induction step. Otherwise, the subgraph $\Gamma(A_n, B_n)$ is colored with at most n different colors. By the above observation it is also ω-good. It thus follows from Theorem 13.6 that in this case there exist a monochromatic subgraph of $\Gamma(A_n, B_n)$ isomorphic to R, which finishes the proof. $\qquad\square$

13.2   Countable-Universal K_ℓ-Free Graphs

In this section we consider subgraphs of Rado's graph which do not contain complete graphs on ℓ vertices.

Definition 13.8. Let $\ell \geq 3$ be a positive integer. By U_ℓ we denote the subgraph of Rado's graph R which is spanned by the vertices $V_\ell = \{n < \omega \mid$ whenever $X \in [n]^{\ell-1}$ spans a $K_{\ell-1}$ in R then there exists $x \in X$ with $\{x, n\} \notin \mathcal{R}\}$.

Obviously, U_ℓ does not contain any complete graph on ℓ vertices, it is K_ℓ-free. Moreover, U_ℓ is universal with respect to the class of all countable K_ℓ-free graphs and its automorphism group acts transitively on finite subgraphs. This is summarized in the next proposition.

Proposition 13.9. *The graph U_ℓ satisfies the following properties:*

(1) *U_ℓ is K_ℓ-free,*
(2) *For any two finite and disjoint sets X and Y in U_ℓ such that X does not contain a complete graph on $(\ell - 1)$ vertices there exists a vertex $z \in V_\ell \setminus (X \cup Y)$ which is joined to all $x \in X$ and not joined to any $y \in Y$.*
(3) *Every countable K_ℓ-free F can be embedded into U_ℓ, moreover, every finite subgraph of U_ℓ which is isomorphic to a subgraph G of F can be extended to an F-subgraph.*
(4) *Any two countable graphs satisfying (1) and (2) are isomorphic.*

Proof. (1) is obvious from the construction, (3) follows from (2) using the same method as in the proof of Proposition 13.2, (4) follows, then, from a back and forth argument. So it remains to show (2). Consider $n_y = 2^{\max Y}$ and $n_x = \sum_{x \in X} 2^x + 2^{n_y}$. By definition of the Rado graph we have

$$\{w < n_x \mid \{w, n_x\} \in \mathcal{R}\} = X \cup \{n_y\}.$$

As n_y is only joined by an edge to $\max Y \notin X$, $X \cup \{n_y\}$ induces no $K_{\ell-1}$ thus we know that $n_x \in V_\ell$.   □

Henson (1971) showed that for every r-coloring of the vertices of U_ℓ, where $\ell \geq 3$ and r is a positive integer, one of the color-classes contains a copy of every finite K_ℓ-free graph. Alternatively, this can also be deduced from Folkman's result (Theorem 12.3). Henson, then, raised the question whether $U_\ell \to (U_\ell)^v_r$. This question was answered positively by El-Zahar and Sauer (1989). Here we only give a proof of the special case when $\ell = 3$ and $r = 2$, which is due to Komjáth and Rödl (1986).

Theorem 13.10.

$$U_3 \xrightarrow{\text{ind}} (U_3)^v_2.$$

Proof. For finite and disjoint subsets $X, Y \subset V_3$, let $\Gamma_3(X, Y) = \Gamma(X, Y) \cap V_3$ denote the set of those vertices in $V_3 \setminus (X \cup Y)$ that are completely connected to X and are not connected to any vertex in Y. From property (2) of Proposition 13.9 we deduce that for every finite subset $Y \subset V_3$ the graph induced by $\Gamma_3(\emptyset, Y)$ is ω-good. As in the proof of Theorem 13.7 it thus follows that the graph induced by $\Gamma_3(\emptyset, Y)$ is isomorphic to U_3.

For a given red-blue coloring of $V_3 = \{v_0, v_1, \ldots\}$, we may assume that both colors appear infinitely many times, as otherwise the previous observation implies that there exists a monochromatic subgraph isomorphic to U_3. Furthermore, assume that there is no red subgraph isomorphic to U_3. We show that this implies the existence of a blue subgraph isomorphic to U_3.

In order to see this we inductively define a sequence of vertices $z_0, z_1, \ldots \in V_3$ and sequences $Y_0, Y_1, \ldots \subset V_3$, $S_0, S_1, \ldots \subset V_3$ and $A_0, A_1, \ldots \subset V_3$ such that the following properties hold for all $n < \omega$:

(1) $S_n = \bigcup_{j<n}(Y_j \cup \{z_j\})$, $Y_n \cap S_n = \emptyset$, and all vertices in Y_i are colored red,
(2) The subgraph spanned by Y_n is isomorphic to the subgraph spanned by $\{v_0, \ldots, v_{k_n}\}$ for some $0 < k_n < \omega$,
(3) $A_n \subseteq Y_n$ such that $A_n = \{j \le k_n \mid \{v_j, v_{k_n+1}\} \in E(U_3)\}$,
(4) Y_n is a maximal subset of V_3 (maximal by set inclusion) with respect to properties (1)–(3),
(5) $z_n \notin S_n \cup Y_n$, and z_n is colored blue,
(6) Let $B_n := \{z_j < n \mid \{v_j, v_n\} \in E(U_3)\}$; if $B_n = \emptyset$ then z_n is not joined to any vertex in $S_n \cup Y_n$, otherwise z_n is joined to all vertices in $B_n \cup A_{n_0}$ and to no vertex in $(S_n \cup Y_n) \setminus (B_n \cup A_{n_0})$, where $n_0 = \min\{j \mid z_j \in B_n\}$.

Clearly, properties (5) and (6) imply that for all $n < \omega$ the subgraph spanned by $\{z_0, \ldots, z_n\}$ is monochromatic (in blue) and isomorphic to the subgraph induced by $\{v_0, \ldots, v_n\}$. Therefore, if we can show that such an infinite sequence exists this will finish the proof.

Assume that we have found a family of subsets $Y_0, \ldots, Y_{n-1}$ and a set of vertices $z_0, \ldots, z_{n-1}$ which satisfy (1)–(6). In order to construct Y_n start with $Y_n = \{v\}$, where v is any red vertex such that $v \notin S_n$ (which exists as we have infinitely many vertices that are colored red). Then greedily add vertices to Y_n so that (2) and (3) remain satisfied. As we assumed that there exits no red monochromatic subgraph isomorphic to U_3, this process will stop with a finite set Y_n satisfying (2)–(4).

If v_n is not joined by an edge to any $v_0, \ldots, v_{n-1}$, then the fact that $\Gamma_3(\emptyset, Y_n \cup S_n)$ is isomorphic to U_3 implies that it cannot contain only red vertices colored; thus taking z_n to be any blue vertex in $\Gamma_3(\emptyset, Y_n \cup S_n)$ suffices.

Otherwise, let B_n be as defined in (6) and let $n_0 := \min\{j \mid z_j \in B_n\}$. If $\Gamma_3(B_n \cup A_{n_0}, (S_n \cup Y_n) \setminus (B_n \cup A_{n_0}))$ is not empty, then by maximality of Y_{n_0} and the definition of A_{n_0} it contains only vertices colored with blue, and it can easily be seen any such vertex can be taken as z_n, satisfying properties (5) and (6). Therefore it only remains to argue that $\Gamma_3(B_n \cup A_{n_0}, (S_n \cup Y_n) \setminus (B_n \cup A_{n_0}))$ cannot be the empty set.

Observe that, by property (2) of Proposition 13.9, it suffices to show that there is no edge in the subgraph induced by $B_n \cup A_{n_0}$. By the definition of A_{n_0} and the fact that U_3 is K_3-free we know that there exists no edge in the subgraph induced by A_{n_0}. By the definition go B_n we know that any edge between two vertices $x, y \in B_n$ spans a triangle with V_n, which can't be. Finally, assume there is an edge between some vertex $x \in A_{n_0}$ and a vertex $z_i \in B_n$, for some $n_0 < i < n$. Observe that (6) implies that the only case that this can happen is when $i_0 = n_0$. But then we have an edge joining v_{i_0} and v_i, again closing a triangle with v_n. Therefore there is no edge in the subgraph induced by $B_n \cup A_{n_0}$ and we can thus find a vertex z_n which satisfies all properties. □

13.3 Colorings of Edges

Considering colorings of edges it turns out that Rado's graph no longer has the property to arrow itself. In fact an even stronger negative result is known (Erdős et al. 1975).

Proposition 13.11. *Let $K_{\omega,\omega}$ be the complete bipartite graph with both parts being countably infinite. Then there exists a 2-coloring of the edges of Rado's graph R such that no induced $K_{\omega,\omega}$-subgraph is monochromatic. In other words,*

$$R \stackrel{ind}{\nrightarrow} (K_{\omega,\omega})_2^e$$

Proof. We first have to define a 2-coloring of the edges of $R = (\omega, \mathcal{R})$. The idea is to play with two different orders on $\mathcal{R}$. The first one, denoted by $\leq$, is the usual order of nonnegative integers. To define the second one we need some preparation. Recall that nonnegative integers $k < m$ are joined by an edge if and only if 2^k occurs in the binary expansion of m. That is, if consider then binary expansion of m, i.e.

$$m = \sum_{i \geq 0} m_i 2^i \qquad \text{with } m_i \in \{0, 1\},$$

then all but finitely many of the m_i's are zero and we have $\{k, m\} \in \mathcal{R}$ if and only if $k < m$ and $m_k = 1$.

The second order, denoted by $\preceq$, is the lexicographic order of the binary expansions, from left to right with $0 < 1$. So for $m = \sum_{i \geq 0} 2^i m_i$ and $n = \sum_{i \geq 0} 2^i n_i$ we have $m \preceq n$ if and only if there exists j such that $m_i = n_i$ for all $i < j$ and $m_j < n_j$. Observe that this implies, for example, that all even nonnegative integers precede the odd ones. In general, $\preceq$ measures which of the two numbers is more 'odd' than the other, and this is, then, the larger one.

Now color an edge $\{k, m\} \in \mathcal{R}$ with color 0 if $\leq$ and $\preceq$ coincide on this edge, i.e., $k \leq m$ and $k \preceq m$, and color it with color 1 otherwise.

Fig. 13.1 There is no
monochromatic $K_{\omega,\omega}$

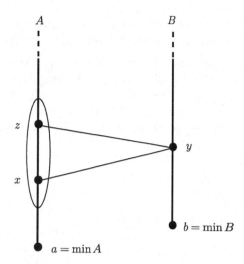

Assume that R contains an induced $K_{\omega,\omega}$-subgraph which is monochromatic. Say, A and B are the two stable parts with $a = \min A < b = \min B$ (Fig. 13.1).

Let A' be an infinite subset of A such that $m_i = n_i$ for all $m, n \in A'$ and all $i \leq a$. Then $m_a = n_a = 0$ as a and m, resp., a and n are not joined by an edge. As A' and B are both infinite there exist $x, z \in A'$ and $y \in B$ such that $x < y < z$. Note that $y_a = 1 > 0 = x_a = z_a$ as a and y are joined by an edge and $a < y$.

If the subgraph is monochromatic in color 0, then the two orders coincide and we have $x \preceq y \preceq z$. As $y \preceq z$ there exists an i such that $y_j = z_j$ for all $j < i$ and $y_i < z_i$. As $y_a > z_a$ this implies $i < a$ and thus, by the definition of A', $y \preceq x$, a contradiction. If the subgraph is monochromatic in color 1, then $z \preceq y \preceq x$. In this case we obtain the desired contradiction similarly as above, with the rôles of x and z interchanged. $\qquad\square$

Corollary 13.12.

$$R \overset{ind}{\nrightarrow} (R)_2^e$$

$\square$

Let us call a graph G locally finite if each vertex of G is joined by an edge only to finitely many vertices of G or, alternatively, it is joined to almost all vertices of G (both kinds of vertices are allowed to occur). Clearly, $K_{\omega,\omega}$ is not locally finite. In contrast to Proposition 13.11, Erdős et al. (1975) prove the following positive partition theorem:

Theorem 13.13. *Let r be a positive integer and let G be a countable locally finite graph. Then for every r-coloring of the edges of Rado's graph there exists* *a monochromatic induced G-subgraph, in other words, $R \overset{ind}{\to} (G)_r^e$ for all positive integers r and countable locally finite graphs G.*   □

There is still a gap between Proposition 13.11 and Theorem 13.13. A characterization of all those countable graphs G satisfying $R \to (G)_r^e$ for every positive integer r is not known. Clearly, such a G must not contain an infinite stable set which is completely joined to another infinite set.

We do not prove Theorem 13.13 here, but refer the reader to Erdős et al. (1975).

Chapter 14
Hypergraphs on Parameter Sets

So far in this chapter, we have studied graphs which are defined on sets. Now we start studying (hyper)graphs which are defined on more complex structures. In particular, in this section we study hypergraphs on parameter sets.

In Sect. 14.1 we prove an induced version of Hales-Jewett's theorem and, as corollaries, we obtain results for sets of integers carrying an arithmetic structure like, e.g., arithmetic progressions or (m, p, c)-sets. In Sect. 14.2 we give an alternative proof of the Ramsey theorem for finite ordered graphs (Theorem 12.13). Though it doesn't exactly fit the theme of the section, it will serve us as a motivating example for a technique which we will then use in Sect. 14.3 to prove an induced version of the Graham-Rothschild's theorem on parameter sets. The induced Graham-Rothschild's theorem gives, in a sense, a complete analogue of the Ramsey theorem for finite ordered graphs.

Before we state these results, we first fix some notation. Given an alphabet A and integers k and n we build an (ordered) hypergraph $\mathcal{H}^k(n)$ as follows. The vertices are all words of length n over A, i.e. $V(\mathcal{H}^k(n)) = A^n$. The set of edges is given by all i-parameter words in $[A]\binom{n}{i}$, for all $0 \leq i \leq k$. More precisely, every $f \in [A]\binom{n}{i}$ corresponds to a hyperedge e_f given by

$$e_f = \{f \cdot g \mid g \in A^i\},$$

and

$$E(\mathcal{H}^k(n)) = \bigcup_{\substack{0 \leq i \leq k \\ f \in [A]\binom{n}{i}}} e_f.$$

Note that we do allow $i = 0$ in the above definition, i.e., all vertices of $\mathcal{H}^k(n)$ are also considered as edges. In this section we will mostly be concerned with finding an appropriate subgraph $\mathcal{F}$ of $\mathcal{H}^k(n)$ that has some nice Ramsey properties. It is important to note that here we do consider weak subgraphs. That is, a subgraph $\mathcal{F} \subseteq \mathcal{H}^k(n)$ can have the property that some vertices of $V(\mathcal{F})$ do not belong to $E(\mathcal{F})$.

H.J. Prömel, *Ramsey Theory for Discrete Structures*,
DOI 10.1007/978-3-319-01315-2_14,

Similarly as in the graph case, for a hypergraph $\mathcal{F} \subseteq \mathcal{H}^k(n)$ and a subset $\mathcal{A} \subseteq A^n$, we denote by $\mathcal{F}[\mathcal{A}]$ the subgraph of $\mathcal{F}$ induced by $\mathcal{A}$, i.e., the vertex set $V(\mathcal{F}[\mathcal{A}])$ is given by $V(\mathcal{F}) \cap \mathcal{A}$ and for all $e_f \in E(\mathcal{F})$ we have

$$e_f \in \mathcal{F}[\mathcal{A}] \quad \text{if and only if} \quad e_f \subseteq \mathcal{A}.$$

We will mostly be interested in subgraphs induced by an m-subspace of A^n, i.e., by some $f \in [A]\binom{n}{m}$. To shorten notation we use $\mathcal{F}[f]$ to denote the subgraph induced by such an m-space:

$$\mathcal{F}[f] := \mathcal{F}[\{f \cdot g \mid g \in A^m\}].$$

14.1   An Induced Hales-Jewett Theorem

For this section let A be a finite set containing at least two elements. As Hales-Jewett's theorem itself, induced versions of Hales-Jewett's theorem consider colorings of A^n, i.e., of vertices. Without loss of generality, we restrict to colorings of vertices which exist as hyperedges.

More precisely, for hypergraphs $\mathcal{E} \subseteq \mathcal{H}^k(m)$ and $\mathcal{F} \subseteq \mathcal{H}^k(n)$, let the Ramsey arrow $\mathcal{F} \to (\mathcal{E})_r^0$ abbreviate the following statement:
For every coloring $\Delta : A^n \to r$ there exists $f \in [A]\binom{n}{m}$ such that $\mathcal{F}[f]$ is isomorphic to $\mathcal{E}$ and $\Delta(f \cdot y) = \Delta(f \cdot x)$ for all $x, y \in A^m$ with $e_x, e_y \in E(\mathcal{E})$.

Note that we require monochromaticity only for those vertices that form an edge in $\mathcal{E}$. Clearly, if all vertices form edges then we get monochromaticity in the usual sense. It is an easy observation that $\mathcal{F}[f]$ is isomorphic to $\mathcal{E}$ if and only if for every $g \in [A]\binom{m}{i}$, $i \leq k$, we have $e_{f \cdot g} \in E(\mathcal{F})$ iff $e_g \in E(\mathcal{E})$. Note that this condition needs to hold for all edges, also those which form vertices.

With this notation at hand, we can state the induced version of Hales-Jewett's theorem:

Theorem 14.1 (Induced Hales-Jewett theorem). *Let r, m and k be positive integers and let $\mathcal{E} \subseteq \mathcal{H}^k(m)$ be given. Then there exists a positive integer n and a subgraph $\mathcal{F} \subseteq \mathcal{H}^k(n)$ such that $\mathcal{F} \to (\mathcal{E})_r^0$.*

Recall that with respect to ordinary graphs the corresponding vertex partition theorem can be established using a simple product construction (cf. Sect. 12.1). Essentially the same idea applies here.

Convention. Recall that in Sect. 4.1 we introduced $\times$ to concatenate two parameter words. In order to get a subspace whose dimension is the sum of the two subspaces we there shifted the parameters in the second word. In this section we only need the *formal* concatenation of two parameter words. With abuse of notation we thus let $\times$ denote in this section the formal concatenation, i.e., for $g = (g_0, \ldots, g_{m-1}) \in [A]\binom{m}{k}$ and $h = (h_0, \ldots, h_{\tilde{m}-1}) \in [A]\binom{\tilde{m}}{i}$ with $i \leq k$ we let

$g \times h = (g_0, \ldots, g_{m-1}, h_0, \ldots, h_{\tilde{m}-1}) \in [A]\binom{m+\tilde{m}}{k}$. In this section we will mostly be concerned with a product space given by the concatenations of a set of parameter words. More precisely, for $B \subseteq \bigcup_{i \leq k} [A]\binom{m}{i}$ let $(B)_M = \{f_0 \times \ldots \times f_{M-1} \mid f_i \in B\}$ which, then, is a subset of $\bigcup_{i \leq k} [A]\binom{m \cdot M}{i}$.

Proof of Theorem 14.1. Consider the set $B = \{x \in A^m \mid e_x \in E(\mathcal{E})\}$. According to Hales-Jewett's theorem (Theorem 4.2) let the positive integer N be such that $N \geq HJ(|B|, 1, r)$. Let $n = N \cdot m$.

We define $\mathcal{F} \subseteq \mathcal{H}^k(n)$ as follows. The vertex set of $\mathcal{F}$ is $V(\mathcal{F}) = A^n$, i.e., it is identical to that of $\mathcal{H}^k(n)$. For $g \in [A]\binom{m}{i}$ such that $e_g \in E(\mathcal{E})$, add $e_{\tilde{g}}$ to $E(\mathcal{F})$ for all $\tilde{g} \in (B \cup \{g\})_N$ such that $\tilde{g}_j = g$ for some $j < N$. Note that in this case $\tilde{g} = (g_0, \ldots, g_{N-1})$ is an element of $[A]\binom{n}{i}$. It remains to verify that $\mathcal{F} \to (\mathcal{E})_r^0$.

Let $\Delta : A^n \to r$ be a coloring. As $(B)_N \subseteq A^n$, by abuse of language this can be viewed as a coloring $\Delta : B^N \to r$. By choice of N there exists a one-parameter word $\tilde{f} \in [B]\binom{N}{1}$ such that the set $\{\tilde{f} \cdot x \mid x \in B\}$ is monochromatic with respect to Δ. Consider an m-parameter word $f \in [A]\binom{n}{m}$ defined as $f = \tilde{f}_0^* \times \ldots \times \tilde{f}_{N-1}^*$ where

$$\tilde{f}_i^* = \begin{cases} \tilde{f}_i, & \text{if } \tilde{f}_i \in B, \\ (\lambda_0, \ldots, \lambda_{m-1}), & \text{if } \tilde{f}_i = \lambda_0. \end{cases}$$

It is clear from the construction of f that $\Delta(f \cdot x) = \Delta(f \cdot y)$ for every $x, y \in A^m$ such that $e_x, e_y \in E(\mathcal{E})$. Moreover, for every $g \in [A]\binom{m}{i}$ we have $f \cdot g \in (B \cup \{g\})_N$, thus $e_{f \cdot g} \in E(\mathcal{F})$ iff $e_g \in E(\mathcal{E})$ and so $\mathcal{F}[f]$ is isomorphic to $\mathcal{E}$. $\square$

14.1.1   Applications

Apparently (Spencer 1975b) first considered induced partition theorems for other structures than graphs defined on sets, by proving an induced version of van der Waerden's theorem on arithmetic progressions. We have seen in Sect. 4.2.1 that van der Waerden's theorem on arithmetic progressions can be easily deduced from Hales-Jewett's theorem. Basically following the lines of this proof we show how an induced version of van der Waerden's theorem can be deduced from Theorem 14.1.

Theorem 14.2 (Induced van der Waerden). *Let r and m be positive integers and let $\mathcal{E} = (m, E)$ be a hypergraph on the vertex set m. Then there exists a positive integer n and a hypergraph $\mathcal{F} = (n, F)$ on the vertex set n, such that for every r-coloring $\Delta : n \to r$ there exists an arithmetic progression $A = \{a + j \cdot b \mid 0 \leq j < m\} \subseteq n$ such that*

(1) *The subgraph of $\mathcal{F}$ spanned by A is isomorphic to $\mathcal{E}$, and*
(2) *$\Delta \rceil \{a + j \cdot b \mid j < m \text{ and } j \in E\}$ is a constant coloring.*

Remark 14.3. Observe that Theorem 14.2 generalizes the particular case of vertex colorings from the Ramsey theorem for ordered graphs in a somewhat unexpected direction. Considering hypergraphs whose vertex sets are integers (i.e., carry an arithmetic structure) the additional requirement is that the vertex set of the monochromatic hypergraph forms an arithmetic progression.

Proof of Theorem 14.2. Let the positive integer r and the hypergraph $\mathcal{E} = (m, E)$ be given. Let $A = m$ and consider the hypergraph $\mathcal{E}_0 \subseteq \mathcal{H}^0(1)$ such that $i \in E(\mathcal{E}_0)$ if and only if $i \in E(\mathcal{E})$, for $i < m$. Now we apply Theorem 14.1 and find a positive integer n_0 and a hypergraph $\mathcal{F}_0 \subseteq \mathcal{H}^0(n_0)$ such that $\mathcal{F}_0 \to (\mathcal{E}_0)_r^0$. Let $n = m^{n_0}$ and recall that $\mathcal{P}(n)$ denotes the power set of n. We define the required hypergraph $\mathcal{F}$ with vertex set $n = \{0, \ldots, n-1\}$ and edges $E(\mathcal{F}) \subseteq \mathcal{P}(n)$ as follows.

Let $\varphi : A^{n_0} \to n$ be such that $\varphi(a_0, \ldots, a_{n_0-1}) = \sum_{i<n_0} a_i \cdot m^i$. Note that φ is a bijection. For every $(a_0, \ldots, a_{n_0-1}) \in A^{n_0}$ let $\varphi(a_0, \ldots, a_{n_0}) \in E(\mathcal{F})$ iff $(a_0, \ldots, a_{n_0-1}) \in E(\mathcal{F}_0)$. Furthermore, for every $f \in [A]\binom{n_0}{1}$ and $J \in \mathcal{P}(m)$, $|J| \geq 2$, let

$$\{\varphi(f \cdot j) \mid j \in J\} \in E(\mathcal{F}) \quad \text{if and only if} \quad J \in E(\mathcal{E}),$$

where $f \cdot j$ refers to composition of parameter words. Observe that $\mathcal{F}$ is well-defined since any two distinct one-parameter sets intersect in at most one point and the mapping φ is a bijection. It remains to verify that the hypergraph $\mathcal{F}$ has the desired properties.

Let $\Delta : n \to r$ be an r-coloring. This defines a coloring $\Delta^* : A^{n_0} \to r$ by $\Delta^*(a_0, \ldots, a_{n_0-1}) = \Delta(\sum_{i<n_0} a_i \cdot m^i)$. By choice of the parameter-graph $\mathcal{F}_0$ there exists $f \in [A]\binom{n_0}{1}$ such that $\mathcal{E}_0$ is isomorphic to $\mathcal{F}_0[f]$ and

$$\Delta^*{\restriction}\{f \cdot j \mid j < m \text{ and } j \in E(\mathcal{E}_0)\} = \Delta{\restriction}\{\varphi(f \cdot j) \mid j < m \text{ and } j \in E(\mathcal{E})\}$$

is a constant coloring. By construction, then, the arithmetic progression $A = \{\varphi(f \cdot j) \mid j < m\}$ has the desired properties.   □

Note that in the above proof the induced version of Hales-Jewett is only applied to the subhypergraph of $\mathcal{E}$ that contains exactly all singleton edges. For the case that all vertices of $\mathcal{E}$ do form an edge one easily checks that the use of the induced Hales-Jewett theorem can be replaced by applying just the classical Hales-Jewett theorem.

Recall that a subset $M \subseteq \mathbb{Z}$ is an (m, p, c)-set if there exist integers $x_0, \ldots, x_m$ such that $M = M_{p,c}(x_0, \ldots, x_m) = \{cx_i + \sum_{j=i+1}^{m} \xi_j x_j \mid -p \leq \xi_j \leq p, \xi_j \in \mathbb{Z}$ for $j = 1, \ldots, m\}$. As seen in Chap. 2, (m, p, c)-sets are a basic tool in studying partition regular systems of equations. Thereby, arithmetic progressions can be viewed as special (m, p, c)-sets, in fact as $(1, p, 1)$-sets. Extending the method of proof used for the induced van der Waerden theorem, Deuber et al. (1982) proved an induced partition theorem for (m, p, c)-sets.

Theorem 14.4. *Let m, p, c and r be positive integers and let (M, E) be a hyper-graph on the set $M = M_{p,c}(x_0, \ldots, x_m)$. Then there exist positive integers n, q, d and there exists a hypergraph (N, F) on the set $N = M_{q,d}(x_0, \ldots, x_n)$ such that for every r-coloring $\Delta : N \to r$ there exists an (m, p, c)-subset $M' \subseteq N$ such that the subgraph of (N, F) spanned by M' is isomorphic to (M, E) and such that $\Delta\rceil\{x \in M' \mid x \in F\}$ is a constant coloring.*   □

The proof basically combines ideas from the proof of the (non induced) partition theorem for (m, p, c)-set (cf. Sect. 4.2.3) and the induced Hales-Jewett resp. van der Waerden theorem. We omit this proof.

14.2 Colorings of Subgraphs: An Alternative Proof

We now reprove Theorem 12.13, the Ramsey theorem for ordered graphs. Instead of using a powerful tool like the Graham-Rothschild theorem for parameter sets (as we did in Sect. 12.3), we now give an elementary proof that uses only Ramsey's theorem and a clever construction. This proof is due to Prömel and Voigt (1989). Recall that the Ramsey theorem for ordered graphs states that for any two ordered two finite graphs $(H, \leq)$ and $(F, \leq)$ and any positive integer r there exists a finite ordered graph $(G, \leq)$ such that

$$(G, \leq) \overset{\text{ind}}{\to} (F, \leq)_r^{(H, \leq)}.$$

Throughout the remainder of this section we assume that all graphs are supplied with an underlying vertex ordering, and that all embeddings and subgraphs respect this ordering, but for ease of notation we will not state these orderings explicitly. In this section the term 'subgraph' also always refers to an *induced* subgraph. In particular, we only color H-subgraphs that are induced H-copies.

Let us first give a high-level overview of our proof strategy. Instead of look-ing directly for an F-subgraph in G which is monochromatic with respect to H-subgraphs, we define another graph F_0. We want that, roughly speaking, F_0 has the following property: if there exists an F_0-subgraph such that the coloring of its H-subgraphs satisfy a certain condition which is, this is the crucial point, much weaker than being monochromatic, then we are guaranteed to find a monochromatic F-subgraph in F_0. Additionally, F_0 will have a strong structural property, namely it is *partite*, which, as we will see, conveniently allows us to find a desired F_0-subgraph iteratively.

14.2.1 Partite Graphs

As usual in graph theory, we say that a graph is *m-partite* if its vertex set can be split into m mutually disjoint and nonempty sets, each inducing an independent set. We impose another strong structural property, namely that it is *left-rectified*.

Definition 14.5. A left-rectified m-partite graph is a pair $((V_v)_{v<m}, E)$, where $V = \bigcup_{i<m} V_i$ is the set of vertices (we assume that the sets V_i are nonempty and mutually disjoint) and

(1) Each V_i induces an independent set, i.e., no edge has both endpoints in the same set V_i,
(2) If $a \in V_{i'}$ and $b \in V_i$ for $i' < i$, then $a \leq b$,
(3) If $\{a, b\} \in E$ for some $a \leq b$ and $a \in V_i$, then $\{a', b\} \in E$ for every $a' \in V$.

Henceforth, we will also call the sets V_i the *parts* of the partition $V = \bigcup_{i<m} V_i$.

Naturally, we want that embeddings of partite graphs preserve the ordering of vertices as well as respect partitions.

Definition 14.6. Let $G = ((V_v)_{v<m}, E)$ and $F = ((\tilde{V}_v)_{v<\tilde{m}})$ be partite graphs. We call a subgraph F of G a *partite F-subgraph of G* if it satisfies the following three conditions: (i) $G[V(F)]$ is isomorphic to F, i.e., F is an *induced F-subgraph of G*, (ii) every part of F is a subset of some part of G and (iii) no two parts of F are subsets of the same part of G. By $\binom{G}{F}_{part}$ we denote the set of all partite F-subgraphs of G.

We say that an m-partite graph is *crossing* if $|V_v| = 1$ for every $v < m$. Note that every graph on m vertices can be viewed as a crossing m-partite graph. Note also that a crossing m-partite graph can easily made left-rectified by ordering the parts in such a way that (2) is satisfied.

Lemma 14.7 (Partite lemma). *Let F and H be left-rectified m-partite graphs with H being crossing, and let r be a positive integer. Then there exists a left-rectified m-partite graph G such that $G \overset{part}{\rightarrow} (F)^H_r$, meaning that for every coloring $\Delta :$ $\binom{G}{H}_{part} \rightarrow r$ there exists a $F \in \binom{G}{F}_{part}$ such that $\Delta \rceil \binom{F}{H}_{part}$ is a constant coloring.*

Proof. We proceed by induction on m. For $m = 1$ the statement reduces to the pigeonhole principle. We prove it for $m + 1$.

Let $F = ((V_v)_{v<m+1}, E_F)$ and $H = (m+1, E_H)$ be $(m+1)$-partite left-rectified graphs where H is crossing, and let r be a positive integer. As H is crossing we may assume that it has $m + 1$ as the set of vertices with parts $\{i\}$ for $i < m + 1$.

Since F is left-rectified, any two vertices $x, x' \in V_m$ which belong to an H-subgraph have the same "profile", i.e., for any $a \in \bigcup_{i<m} V_i$ we have $\{a, x\} \in E_F$ if and only $\{a, x'\} \in E_F$. Let $V_H \subseteq V_m$ be the set of all vertices in V_m which belong to an H-subgraph, and set $z = |V_H|$. Furthermore, let H' and F' be subgraphs of H, resp. F, spanned by the first m parts.

By the induction hypothesis, there exists an m-partite graph G' such that $G' \overset{part}{\rightarrow} (F')^{H'}_{rz^*}$, where $z^* = r \cdot (z - 1) + 1$. Now we extend G' by a set X_m to an $(m + 1)$-partite graph G as follows. First we add vertices $y_0, \ldots, y_{z^*-1}$ to X_m, such that they respect the property of being left-rectified and they all form an H-subgraph with the vertices from G'. Secondly, for each vertex $\hat{x} \in V_m \setminus V_H$ add a vertex to X_m and connect it to the parts in G' in exactly the same way as $\hat{x}$ is connected to the parts $V_i, i < m$ in F. Note that this guarantees that every z-element subset of

Fig. 14.1 The
$*_J$-amalgamation

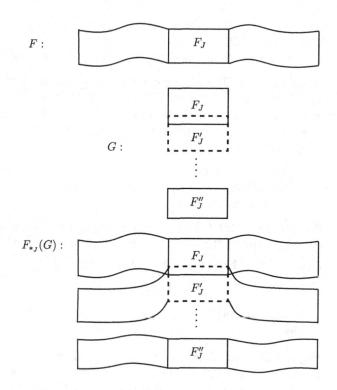

$y_0, \ldots, y_{z^*-1}$ can be extended to a copy of F in G. We claim that the so constructed graph G has the desired properties.

Let $\Delta : \binom{G}{H}_{part} \to r$ be an r-coloring. This induces an r^{z^*} coloring $\Delta^* :$ $\binom{G'}{H'}_{part} \to r^{z^*}$ by $\Delta^*(\tilde{H}') = \langle \Delta(\tilde{H}' \cup \{y_i\}) \mid i < z^* \rangle$. Let $\tilde{G}' \in \binom{F'}{G'}_{part}$ be monochromatic with respect to Δ^*. This induces an r-coloring of the vertices $\{y_0, \ldots, y_{z^*-1}\}$ and by choice of z^* and the pigeonhole principle there exist z of them in the same color. Extending $\tilde{G}'$ with such z vertices and the corresponding $\hat{x}$ vertices yields a partite F-subgraph monochromatic with respect to Δ. □

14.2.2 Amalgamation of Partite Graphs

Having the partite lemma available, we explain our second tool, the $*_J$-amalgamation.

Let $F = ((X_v)_{v<m}, E_F)$ be a left-rectified m-partite graph and let $J \subset m$ be a nonempty subset. By F_J we denote the subgraph of F spanned by the parts $X_j, j \in J$. Additionally, let $G = ((Y_v)_{v \in J}, E_G)$ be a left-rectified $|J|$-partite graph that contains many partite F_J-subgraphs. The idea of the $*_J$-amalgamation is to extend every partite F_J-subgraph of G to an F-graph in a vertex disjoint way, cf. Fig. 14.1.

Finally, we add edges (as few as possible) to ensure that the newly constructed graph is again left-rectified.

Formally, we define the amalgamation $F_{*_J}(G)$ of F with G along F_J as follows:

Definition 14.8. The subgraph of the amalgamation which is spanned by the parts $j \in J$ is precisely G, i.e., $(F_{*_J}(G))_J = G$. Moreover, every $\tilde{F}_J \in \binom{G}{F_J}$ extends to an m-partite graph isomorphic to F such that every two such graphs are mutually disjoint up to the intersection in $(F_{*_J}(G))_J$. The graph $F_{*_J}(G)$ is m-partite and left-rectified.

A moment of thought reveals that such a graph can indeed be constructed. For our need the following property, which can easily be seen to follow from the definitions, is of importance.

Property 14.9. Let F be an m-partite left-rectified graph and $J \subset m$. Let H and G be $|J|$-partite left-rectified graphs, where in addition H is crossing, and assume $G \xrightarrow{\text{part}} (F_J)_r^H$. Then for every r-coloring $\Delta : \binom{F_{*_J}(G)}{H}_{part} \to r$ there exists an $\tilde{F} \in \binom{F_{*_J}(G)}{F}_{part}$ such that $\Delta \rceil \binom{\tilde{F}_J}{H}_{part}$ is a constant coloring.

With these tools at hand, we can now reprove the Ramsey theorem for ordered graphs.

Proof of Theorem 12.13. Let F and H be given graphs. As observed earlier, we can treat them as m-partite, resp. k-partite graphs, where m and k are the number of vertices of F, resp. H. According to Ramsey's theorem let n be such that $n \to (m)_r^k$.

Instead of looking directly for a monochromatic F-subgraph, we define a left-rectified n-partite graph F_0 such that for every $J \in [n]^m$ there exists a (partite) F-subgraph in the partite subgraph $(F_0)_J$ of F_0 spanned by the parts $j \in J$. Such an F_0 can be obtained straightforwardly by placing the required F-subgraphs vertex disjointly and, eventually, adding edges to make it left-rectified. We aim at finding an F_0-subgraph $\tilde{F}_0$ which satisfies the following coloring property,

($\star$) For all $J \in [n]^k$: all H-subgraphs in $(\tilde{F}_0)_J$ are colored monochromatically, i.e. for all $\tilde{H}, \tilde{H}' \in \binom{(\tilde{F}_0)_J}{H}_{part}$ we have $\Delta(\tilde{H}) = \Delta(\tilde{H}')$.

Note that the existence of such an F_0-subgraph implies, by choice of n, that there exists an F-subgraph $\tilde{F}$ such that $\Delta \rceil \binom{\tilde{F}}{H}_{part}$ is a constant coloring. As F is crossing it follows that $\binom{\tilde{F}}{H}_{part}$ coincides with $\binom{\tilde{F}}{H}$, thus we have found the desired monochromatic F-subgraph.

Next we construct an n-partite left-rectified graph G such that for every coloring $\Delta : \binom{G}{H}_{part} \to r$ there exists an F_0-subgraph which satisfies property ($\star$).

Let $(J_i)_{i<q}$ be an enumeration of $[n]^k$. By Lemma 14.7 (partite lemma) there exists a left-rectified n-partite graph F_0^* such that

$$F_0^* \xrightarrow{\text{part}} ((F_0)_{J_0})_r^H,$$

where $(F_0)_{J_0}$ denotes the subgraph of F_0 spanned by parts $j \in J_0$. Let $F_1 = (F_0)_{*J_0}(F_0^*)$, and observe that by Property 14.9, F_1 contains an F_0-subgraph $\tilde{F}_0$ which satisfies property $(\star)$ when restricted to J_0 instead of all $J \in [n]^k$.

We continue the construction in the same way. Assume that we have constructed a graph F_i such that for any coloring of $\binom{F_i}{H}_{part}$ there exists an F_0-subgraph $\tilde{F}_0$ which satisfies property $(\star)$ when restricted to sets $J_0, \ldots, J_{i-1}$. Then let F_i^* be such that $F_i^* \overset{part}{\to} ((F_i)_{J_i})_r^H$ and set $F_{i+1} = (F_i)_{*J_i}(F_i^*)$. Now we have that for any coloring of partite H-subgraphs of F_{i+1} there exists an F_i-subgraph $\tilde{F}_i$ such that $\Delta\rceil\binom{(\tilde{F}_i)_{J_i}}{H}_{part}$ is a constant coloring. However, such $\tilde{F}_i$ now contains an F_0 subgraph $\tilde{F}_0$ which satisfies property $(\star)$ when restricted to restricted to $J_0, \ldots, J_i$.

Repeating the same argument inductively, we have that for any coloring $\Delta : \binom{F_q}{H}_{part} \to r$ there exists an F_0-subgraph which satisfies property $(\star)$. By the earlier observation, this implies the existence of a monochromatic F-subgraph, thus setting $G = F_q$ proves the theorem.   □

Remark 14.10. The approach presented in this section can be extended to also obtain a *restricted* version of the Ramsey theorem for ordered graphs, cf. Prömel and Voigt (1989). In Chap. 16 we consider restricted Ramsey theorems from a different view point.

14.3   An Induced Graham-Rothschild Theorem

In this section we prove an induced version of the Graham-Rothschild theorem. This generalizes the Graham-Rothschild partition theorem for parameter sets in the same way as the Ramsey's theorem for ordered graphs defined on sets generalizes Ramsey's theorem.

The induced Graham-Rothschild theorem has been proved originally in Prömel (1985). Somewhat simpler proofs, then, have been given in Frankl et al. (1987) and Prömel and Voigt (1988).

Definition 14.11. For hypergraphs $\mathcal{F} \subseteq \mathcal{H}^k(m)$ and $\mathcal{G} \subseteq \mathcal{H}^k(n)$, by $\binom{\mathcal{G}}{\mathcal{F}}$ we denote the set of all m-parameter words $f \in [A]\binom{n}{m}$ such that $\mathcal{G}[f]$ is isomorphic to $\mathcal{F}$.

Theorem 14.12 (Induced Graham-Rothschild theorem). *Let A be an alphabet of size $|A| \geq 2$, k and r be positive integers, and let $\mathcal{F} \subseteq \mathcal{H}^k(m)$ and $\mathcal{E} \subseteq \mathcal{H}^k(t)$ be given hypergraphs. Then there exists a positive integer n and a hypergraph $\mathcal{G} \subseteq \mathcal{H}^k(n)$ such that $\mathcal{G} \to (\mathcal{F})_r^{\mathcal{E}}$, i.e., for every $\Delta : \binom{\mathcal{G}}{\mathcal{E}} \to r$ there exists an $f \in \binom{\mathcal{G}}{\mathcal{F}}$ such that $\Delta\rceil\binom{\mathcal{G}[f]}{\mathcal{E}}$ is a constant coloring.*

The assumption $|A| \geq 2$ is just for convenience. For $|A| = 1$ the proof requires some additional twists, cf. Prömel and Voigt (1988).

Recall that with respect to hypergraph $\mathcal{E} = \mathcal{H}^0(0)$, i.e., the case of vertex colorings, the theorem reduces to the induced Hales-Jewett theorem which has been proved in Sect. 14.1.

As the proof of Theorem 14.12 is quite involved, let us first give a very high-level overview of our proof strategy. In fact, the general approach is very similar to the one that we just saw for the graph case in the previous section. In order to transfer these ideas to the hypergraph case we first need to generalize the notations of 'partiteness' and 'amalgamation' from the graph setting to hypergraphs define on parameters sets. In a second step we will use these notions to define an appropriate hypergraph $\mathcal{F}_0$ (that takes over the rôle of F_0 in the graph case). The structural properties of $\mathcal{F}_0$ will then allows us, again similar as in the graph case, to construct the desired hypergraph $\mathcal{G}$ iteratively.

14.3.1 Partite Hypergraphs

As a first step in the proof of the induce Graham-Rothschild theorem, we define an appropriate notion of 'partiteness'. While we will eventually have the property that the 'parts' are stable (contain no edges), we here use a different approach of defining the 'parts'. Consider $\mathcal{H}^k(m + n)$. Its vertices are words of length $m + n$ over the alphabet A. The idea is to use the first m letters to describe the 'part' and the remaining n letters to describe the vertices within a part. Note that in this way an m-partite graph will actually consist of $|A|^m$ parts. We also want that edges in an m-partite graph are 'crossing', meaning that they contain at most one vertex from each part. We now give a formal definition.

Let $f \in [A]\binom{m+n}{j}$ be a parameter word. We write $\dim f = j$ indicating that f is a j-parameter word. By $f\rceil m$ we denote the restriction of f to the first m entries (coordinates). Recall that, formally, f is a mapping $f : m + n \to A \cup \{\lambda_0, \ldots, \lambda_{j-1}\}$. So the restriction $f\rceil m$ again is a parameter word, this time of length m. Observe that $\dim f\rceil m \leq j$.

Definition 14.13. A hypergraph $\mathcal{E} \subseteq \mathcal{H}^k(m + n)$ is m-partite if $e_g \notin \mathcal{E}$ whenever $\dim g\rceil m < \dim g$. A partite embedding of an m-partite hypergraph $\mathcal{E} \subseteq \mathcal{H}^k(m+n)$ into $\tilde{\mathcal{E}} \subseteq \mathcal{H}^k(m + \tilde{n})$ is given by an $f \in [A]\binom{m+\tilde{n}}{m+n}$ such that $\tilde{\mathcal{E}}[f]$ is isomorphic to $\mathcal{E}$ and $\dim f\rceil m = m$. By $\binom{\tilde{\mathcal{E}}}{\mathcal{E}}_{part}$ we denote the set of partite $\mathcal{E}$-subgraphs of $\tilde{\mathcal{E}}$, i.e., the set of all partite embeddings of $\mathcal{E}$ into $\tilde{\mathcal{E}}$.

With respect to sets A having at least two elements, an m-partite hypergraph $\mathcal{E} \subseteq \mathcal{H}^k(m + n)$ can be visualized as follows. The set of vertices $[A]\binom{m+n}{0}$ is split into sets $x \times [A]\binom{n}{0}, x \in [A]\binom{m}{0}$, which we call the *parts* of $\mathcal{E}$. Then the edges have to be *crossing*, i.e., intersect each partition at most once. In other words, ignoring the hyperedges containing only a single vertex, each partition then forms an independent set. Being crossing is reflected by the requirement that $e_g \in \mathcal{E}$ only if $\dim g\rceil m = \dim g$. In particular every hypergraph $\mathcal{E} \subseteq \mathcal{H}^k(m)$ can be viewed as a (crossing) m-partite hypergraph. Finally, the requirement on partite embeddings ensures that each part of $\mathcal{E}$ is inscribed into some (unique) part of $\tilde{\mathcal{E}}$.

Lemma 14.14 (Partite lemma). *Let $\mathcal{F} \subseteq \mathcal{H}^k(m+n)$ and $\mathcal{E} \subseteq \mathcal{H}^k(m)$ be m-partite hypergraphs and let r be a positive integer. Then there exists a positive integer $\tilde{n}$ and an m-partite hypergraph $\mathcal{G} \subseteq \mathcal{H}^k(m+\tilde{n})$ satisfying $\mathcal{G} \overset{part}{\to} (\mathcal{F})^{\mathcal{E}}_r$, meaning that for every $\Delta : \binom{\mathcal{G}}{\mathcal{E}}_{part} \to r$ there exists a partite embedding $f \in \binom{\mathcal{G}}{\mathcal{F}}_{part}$ such that $\Delta \restriction \binom{\mathcal{G}[f]}{\mathcal{E}}_{part}$ is a constant coloring.*

Proof. The proof of Lemma 14.14 just uses Hales-Jewett's theorem and is somewhat similar to the proof of the induced Hales-Jewett theorem (Theorem 14.1).

Recall that $\binom{\mathcal{F}}{\mathcal{E}}_{part} \subseteq \{f \in [A]\binom{m+n}{m} \mid f \restriction m = (\lambda_0, \ldots, \lambda_{m-1})\}$. In particular we have that $f \restriction m = \tilde{f} \restriction m$ for any two $f, \tilde{f} \in \binom{\mathcal{F}}{\mathcal{E}}$. We cut off the first m entries of each such f and let

$$\mathcal{T} = \{g \in (A \cup \{\lambda_0, \ldots, \lambda_{m-1}\})^n \mid (\lambda_0, \ldots, \lambda_{m-1}) \times g \in \binom{\mathcal{F}}{\mathcal{E}}_{part}\}$$

be the set of tails. Let the positive integer s be such that $s \geq HJ(|\mathcal{T}|, 1, r)$, and consider the set

$$\mathcal{T}^* = \{g_0 \times \ldots \times g_{s-1} \mid g_i \in \mathcal{T} \text{ or } g_i = (\lambda_m, \ldots, \lambda_{m+n-1}) \text{ for all } i < s \text{ and}$$

$$g_j = (\lambda_m, \ldots, \lambda_{m+n-1}) \text{ for at least one } j < s\}.$$

Observe that $\mathcal{T}^*$ corresponds to the set of one-parameter words $[\mathcal{T}]\binom{s}{1}$, where, for convenience, the parameter is replaced by $(\lambda_m, \ldots, \lambda_{m+n-1})$. Also observe that $(\lambda_0, \ldots, \lambda_{m-1}) \times \mathcal{T}^* \subseteq [A]\binom{m+n\cdot s}{m+n}$.

We now define a hypergraph $\mathcal{G} \subseteq \mathcal{H}^k(m + n \cdot s)$. For a $h \in \mathcal{T}^*$ let $\overline{\lambda}h = (\lambda_0, \ldots, \lambda_{m-1}) \times h$. Then for every $h \in \mathcal{T}^*$ and for every $g \in [A]\binom{m+n}{i}$ set

$$e_{\overline{\lambda}h \cdot g} \in \mathcal{G} \text{ if and only if } e_g \in \mathcal{F}.$$

The following claim shows, and this is where the property of being partite comes into play, that $\mathcal{G}$ is well-defined.

Claim. Let $g, g' \in [A]\binom{m+n}{i}$ and let $h, h' \in \mathcal{T}^*$. Assume that $g \neq g'$ and $\overline{\lambda}h \cdot g = \overline{\lambda}h' \cdot g'$. Then $e_g \in \mathcal{F}$ iff $e_{g'} \in \mathcal{F}$.

Proof of Claim. First observe that $\overline{\lambda}h \cdot g = \overline{\lambda}h' \cdot g'$ implies that $g \restriction m = g' \restriction m$, so g and g' differ only in their tail sequence. If $\dim g \restriction m < i$, then by the definition we have $e_g \notin \mathcal{G}$ and $e_{g'} \notin \mathcal{G}$, thus we are done.

Otherwise, let $h = h_0 \times \ldots \times h_{s-1}$ and $h' = h'_0 \times \ldots \times h'_{s-1}$. Since $h \in \mathcal{T}^*$ there exists an $j < s$ such that $h_j = (\lambda_m, \ldots, \lambda_{m+n-1})$. Then $h'_j \in \mathcal{T}$, as otherwise we would have $((\lambda_0, \ldots, \lambda_{m-1}) \times h_j) \cdot g \neq ((\lambda_0, \ldots, \lambda_{m-1}) \times h'_j) \cdot g'$ and so $\overline{\lambda}h \cdot g \neq \overline{\lambda}h' \cdot g'$. Thus $((\lambda_0, \ldots, \lambda_{m-1}) \times h'_j) \cdot (g' \restriction m) = g$. Moreover, as $h'_j \in \mathcal{T}$

we know that $(\lambda_0, \ldots, \lambda_{m-1}) \times h'_j \in \left(\begin{smallmatrix} \mathcal{F} \\ \mathcal{E} \end{smallmatrix}\right)_{part}$, hence $e_{g'\rceil m} \in \mathcal{E}$ iff $e_g \in \mathcal{F}$. Using the same argument we deduce $e_{g\rceil m} \in \mathcal{E}$ iff $e_{g'} \in \mathcal{F}$, which together with the observation $g\rceil m = g'\rceil m$ proves the claim.

It remains to verify that indeed $\mathcal{G} \xrightarrow{part} (\mathcal{F})^{\mathcal{H}}_r$. Let $\Delta : \left(\begin{smallmatrix} \mathcal{G} \\ \mathcal{E} \end{smallmatrix}\right)_{part} \to r$ be an r-coloring. This induces an r-coloring of $[T]\left(\begin{smallmatrix} s \\ 0 \end{smallmatrix}\right)$ and thus, by choice of s, there exists a monochromatic line which can be identified with some $h \in T^*$. Now by the construction of $\mathcal{G}$ we have that $\mathcal{G}[\overline{\lambda h}]$ is isomorphic to $\mathcal{F}$, yielding the desired monochromatic $\mathcal{F}$-subgraph. $\qquad\qquad\square$

14.3.2 Amalgamation of Partite Graphs

In this section we describe the concept of *amalgamation*. Again, we first fix some notation.

Let $\mathcal{F} \subseteq \mathcal{H}^k(m+n)$ be an m-partite hypergraph. Then for $h \in [A]\left(\begin{smallmatrix} m \\ t \end{smallmatrix}\right)$, by $\mathcal{F}_h$ we denote the t-partite hypergraph isomorphic to $\mathcal{F}[h \times (\lambda_t, \ldots, \lambda_{t+n-1})]$, or, more precisely,

$$\mathcal{F}_h = \mathcal{F}[\{h \cdot x \mid x \in \left(\begin{smallmatrix} t \\ 0 \end{smallmatrix}\right)\} \times [A]\left(\begin{smallmatrix} n \\ 0 \end{smallmatrix}\right)].$$

Intuitively, $\mathcal{F}_h$ is a subgraph spanned by a subset of the partition of $\mathcal{F}$ specified by the parameter word h.

Additionally, let $\mathcal{G} \subseteq \mathcal{H}^k(t+\tilde{n})$ be a t-partite hypergraph. The idea of an $*h$-amalgamation is exactly as in the similar notion of a $*_J$-amalgamation in the graph case: we want an m-partite graph $\mathcal{F}_{*h}(\mathcal{G})$ that extends every $\mathcal{F}_h$-subgraph in $\left(\begin{smallmatrix} \mathcal{G} \\ \mathcal{F}_h \end{smallmatrix}\right)_{part}$ to an $\mathcal{F}$-graph in a 'vertex-disjoint way'. For a formal definition let $g_0, \ldots, g_{z-1}$ be an enumeration of the partite $\mathcal{F}_h$-subgraphs in $\mathcal{G}$.

Definition 14.15. A hypergraph $\mathcal{F}_{*h}(\mathcal{G}) \subseteq \mathcal{H}^k(m+n')$ is an $*h$-amalgamation of $\mathcal{F}$ with $\mathcal{G}$ along h if the following holds: $\mathcal{F}_{*h}(\mathcal{G})$ is m-partite and there exist $f_0^*, \ldots, f_{z-1}^* \in \left(\begin{smallmatrix} \mathcal{F}_{*h}(\mathcal{G}) \\ \mathcal{F} \end{smallmatrix}\right)_{part}$ such that the intersection of $\mathcal{F}_{*h}[f_i^*]$ and $\mathcal{F}_{*h}[f_j^*]$ is isomorphic to the intersection of $\mathcal{G}[g_i]$ and $\mathcal{G}[g_j]$. In particular, we require that $(\mathcal{F}_{*h}(\mathcal{G}))_h$ is isomorphic to $\mathcal{G}$.

The next lemma shows that such a hypergraph $\mathcal{F}_{*h}(\mathcal{G})$ indeed exists.

Lemma 14.16. *Let $\mathcal{G} \subseteq \mathcal{H}^k(t+\tilde{n})$ and $\mathcal{F} \subseteq \mathcal{H}^k(m+n)$ be given t-partite, resp. m-partite hypergraphs and let $h \in [A]\left(\begin{smallmatrix} m \\ t \end{smallmatrix}\right)$. Then there exists an $*h$-amalgamation $\mathcal{F}_{*h}(\mathcal{G}) \subseteq \mathcal{H}^k(m+\tilde{n}+(z+1)\cdot m)$ of $\mathcal{F}$ with $\mathcal{G}$ along h, where z denotes the cardinality of $\left(\begin{smallmatrix} \mathcal{G} \\ \mathcal{F}_h \end{smallmatrix}\right)_{part}$.*

As in the graph case the importance of this amalgamation technique stems from its strong coloring properties. The following proposition (that follows immediately from the definition of the amalgamation) captures this feature. This proposition is all we need in the subsequent section for the proof of the induced Graham-Rothschild theorem.

Proposition 14.17 (Coloring property of $*h$-amalgamation). *Let $\mathcal{E} \subseteq \mathcal{H}^k(t)$ be a t-partite hypergraph and assume that $\mathcal{G} \overset{part}{\to} (\mathcal{F}_h)_r^{\mathcal{E}}$. Then for every coloring Δ :*
$\left(\frac{\mathcal{F}_{*h}(\mathcal{G})}{\mathcal{E}}\right)_{part} \overset{part}{\to} r$ *there exists an $f \in \left(\frac{\mathcal{F}_{*h}(\mathcal{G})}{\mathcal{F}}\right)_{part}$ such that for $\tilde{\mathcal{F}} = \mathcal{F}_{*h}(\mathcal{G})[f]$ we have that $\Delta\rceil\left(\frac{\tilde{\mathcal{F}}_h}{\mathcal{E}}\right)_{part}$ is a constant coloring.* □

The remainder of this section is devoted to the (somewhat technical) proof of Lemma 14.16. The first lemma shows that for every $h \in [A]\binom{m}{t}$ and every positive integer z there exist z distinct m-parameter sets in $[A]\binom{(z+1)m}{m}$ which mutually intersect in their h-subspace.

Lemma 14.18. *Let $h \in [A]\binom{m}{t}$ and let z be a positive integer. Then there exist parameter words $f_i \in [A]\binom{(1+z)\cdot m}{m}$ for $i < z$ with the following properties.*

(1) $f_i \cdot x = f_j \cdot x$ *for all $i < j < z$ and all $x \in h \cdot [A]\binom{t}{0}$,*

(2) $f_i \cdot x \neq f_j \cdot x'$ *for all $i < j < z$ and all $x \in [A]\binom{m}{0} \setminus h \cdot [A]\binom{t}{0}$ and all $x' \in [A]\binom{m}{0}$.*

To understand the proof of the lemma properly some familiarity with the formal calculus of parameter words may be helpful. As we slightly extend the composition of parameter words also to non-parameter words let us recall the basic definition.

Let $g = (g_0, \ldots, g_{n-1}) \in (A \cup \{\lambda_0, \ldots, \lambda_{m-1}\})^n$ and let $h = (h_0, \ldots, h_{m-1}) \in (A \cup \{\lambda_0, \ldots, \lambda_{t-1}\})^m$. Note that neither g nor h are required to be parameter words in the sense of Sect. 3.1. Still we define the composition $g \cdot h \in (A \cup \{\lambda_0, \ldots, \lambda_{t-1}\})^n$ straightforwardly, viz., $g \cdot h = (f_0, \ldots, f_{n-1})$ where

$$f_i = \begin{cases} g_i, & \text{if } g_i \in A, \\ h_j, & \text{if } g_i = \lambda_j. \end{cases}$$

Proof of Lemma 14.18. Let $h = (h_0, \ldots, h_{m-1}) \in [A]\binom{m}{t}$. For every $j < t$ we define j' as the minimal index at which λ_j appears: $j' = \min\{i < m \mid h_i = \lambda_j\}$.

Consider $y = (y_0, \ldots, y_{m-1}) \in (A \cup \{\lambda_{j'} \mid j < t\})^m$ which is defined by

$$y_i = \begin{cases} h_i, & \text{if } h_i \in A, \\ \lambda_{j'}, & \text{if } h_i = \lambda_j. \end{cases}$$

We now show that $y \cdot x = x$ if and only if $x \in h \cdot [A]\binom{t}{0}$. Since by construction we have $y \cdot h = h$, it easily follows that $x \in h \cdot [A]\binom{t}{0}$ implies $y \cdot x = x$. On the other hand, y and h have the same pattern: if $h_i = h_j = \lambda_k$ then $y_i = y_j = \lambda_{k'}$. Thus, $y \cdot x = x$ implies $(y \cdot x)_i = h_i = x_i$ if $h_i \in A$ and $(y \cdot x)_i = x_{k'} = x_i$ if $h_i = \lambda_k$. Hence, $x \in h \cdot [A]\binom{t}{0}$.

Now we define $f_i \in [A]\binom{(1+z)\cdot m}{m}$ by

$$f_i = (\lambda_0, \ldots, \lambda_{m-1}) \times \underbrace{y \times \ldots \times y}_{i \text{ times}} \times (\lambda_0, \ldots, \lambda_{m-1}) \times \underbrace{y \times \ldots \times y}_{z-1-i \text{ times}}.$$

As each f_i starts with $(\lambda_0, \ldots, \lambda_{m-1})$ assertion (2) is obviously satisfied for $x \neq x'$. The remaining cases follow from the fact that $y \cdot x = x$ if and only if $x \in h \cdot [A]\binom{t}{0}$.
□

Also the next lemma sounds somewhat technical. Its significance will be clear in the construction of the amalgamation.

The problem is the following: consider the embedding $g_i \in \binom{\mathcal{F}_h}{\mathcal{G}}$, so $g_i \in [A]\binom{t+\tilde{n}}{t+n}$. We want to find a $g_i^* \in [A]\binom{m+\tilde{n}}{m+n}$ such that $g_i^* \cdot (h \times (\lambda_k, \ldots, \lambda_{k+n-1}))$ behaves like g_i. Recall that $h \times (\lambda_k, \ldots, \lambda_{k+n-1}) \in \binom{\mathcal{F}}{\mathcal{F}_h}$.

Lemma 14.19. *Let $g \in [A]\binom{t+\tilde{n}}{t+n}$ be such that $\dim g \rceil t = t$, thus g can be written as $g = (\lambda_0, \ldots, \lambda_{t-1}) \times g_{tail}$. Let $h \in [A]\binom{m}{t}$. Then there exists $g^* \in [A]\binom{m+\tilde{n}}{m+n}$ which can be written as $g^* = (\lambda_0, \ldots, \lambda_{m-1}) \times g_{tail}^*$ such that for all $f \in [A]\binom{t+n}{i}$ it follows that*

$$g^* \cdot ((h \times (\lambda_t, \ldots, \lambda_{t+n-1})) \cdot f) = (h \cdot f \rceil t) \times (g_{tail} \cdot f).$$

Proof. As before, let $j' = \min\{i < m \mid h_i = \lambda_j\}$, for all $j < t$. Let $g_{tail} = (\alpha_0, \ldots, \alpha_{\tilde{n}-1})$. Then setting $g^* = (\lambda_0, \ldots, \lambda_{m-1}) \times (\alpha_0^*, \ldots, \alpha_{\tilde{n}-1}^*)$, where

$$\alpha_i^* = \begin{cases} \alpha_i, & \text{if } \alpha_i \in A \\ \lambda_{m+j}, & \text{if } \alpha_i = \lambda_{t+j}, \\ \lambda_{j'}, & \text{if } \alpha_i = \lambda_j \text{ for } j < t \end{cases}$$

proves the lemma.
□

Now we are in the position to prove Lemma 14.16.

Proof of Lemma 14.16. Let $h \in [A]\binom{m}{t}$ and $(g_i)_{i<z}$ be an enumeration of $\binom{\mathcal{F}_h}{\mathcal{G}}_{part}$. Let the parameter words $f_i \in [A]\binom{(z+1)\cdot m}{m}$ for $i < z$ be as in Lemma 14.18. Also let $g_i^* \in [A]\binom{m+\tilde{n}}{m+n}$ be as in Lemma 14.19 with respect to h and g_i. Now we define $\mathcal{F}_{*h}(\mathcal{G})$ as follows:

$$e_{(g_j^* \times f_j) \cdot g} \in \mathcal{F}_{*h}(\mathcal{G}) \text{ iff } e_g \in \mathcal{F}$$

for all $i \leq k$ and $g \in [A]\binom{m+n}{i}$ and all $j < z$. The following claim shows that this is a proper definition.

Claim. Let $i \leq k$ and $g, g' \in [A]\binom{m+n}{i}$ and let $j < j' < z$ be such that $(g_j^* \times f_j) \cdot g = (g_{j'}^* \times f_{j'}) \cdot g'$. Then $e_g \in \mathcal{F}$ if and only if $e_{g'} \in \mathcal{F}$.

Proof of the Claim. As $g_j^* \rceil m = g_{j'}^* \rceil m = (\lambda_0, \ldots, \lambda_{m-1})$ we see that $g \rceil m = g' \rceil m$. Without loss of generality we can assume that g and g' are crossing, i.e., $g \rceil m \in [A]\binom{m}{i}$. From Lemma 14.18 we conclude that $g \rceil m \in h \cdot [A]\binom{t}{i}$. In other words, there exist $f, f' \in [A]\binom{t+n}{i}$ such that $g = (h \times (\lambda_t, \ldots, \lambda_{m+n-1})) \cdot f$, resp., $g' = (h \times (\lambda_t, \ldots, \lambda_{m+n-1})) \cdot f'$. From Lemma 14.19 it follows that

$$g_j^* \cdot g = (h \cdot f \rceil t) \times (g_{j,tail} \cdot f), \quad \text{resp.,} \quad g_{j'}^* \cdot g' = (h \cdot f \rceil t) \times (g_{j',tail} \cdot f'),$$

where $g_j = (\lambda_0, \ldots, \lambda_{t-1}) \times g_{j,tail}$ and $g_{j'} = (\lambda_0, \ldots, \lambda_{t-1}) \times g_{j',tail}$.

It follows from $g_j^* \cdot g = g_j^*$ that $g_j \cdot f = g_{j'} \cdot f = g_{j'} \cdot f'$, hence $e_{g_j \cdot f} \in \mathcal{G}$ iff $e_{g_{j'} \cdot f'} \in \mathcal{G}$. On the other hand, as $h \times (\lambda_t, \ldots, \lambda_{t+n-1}) \in \binom{\mathcal{F}}{\mathcal{F}_h}$ and $g_j, g_{j'} \in \binom{\mathcal{F}_h}{\mathcal{G}}$, we see that

$$e_g \in \mathcal{F} \Leftrightarrow e_{(h \times (\lambda_t, \ldots, \lambda_{t+n-1})) \cdot f} \in \mathcal{F}$$
$$\Leftrightarrow e_f \in \mathcal{F}_h \Leftrightarrow e_{g_j \cdot g} \in \mathcal{G} \Leftrightarrow e_{g_{j'} \cdot f'} \in \mathcal{G} \Leftrightarrow e_{f'} \in \mathcal{F}_h$$
$$\Leftrightarrow e_{(h \times (\lambda_t, \ldots, \lambda_{t+n-1})) \cdot f'} \in \mathcal{F} \Leftrightarrow e_{g'} \in \mathcal{F},$$

as desired. □

14.3.3 *Proof of the Induced Graham-Rothschild Theorem*

With these tools at hand, namely induced Graham-Rothschild theorem for partite graphs (Lemma 14.14) and the notion of an $*h$-amalgamation, we can now prove the induced Graham-Rothschild theorem. Actually, the proof is very similar to the one for the ordered Ramsey theorem from the previous section. First we define an appropriate hypergraph $\mathcal{F}_0$ that will allow us to always find the desired monochromatic $\mathcal{F}$-subgraph. In order to construct $\mathcal{F}_0$ we use now the Graham-Rothschild theorem (Theorem 5.1) instead of the classical Ramsey theorem. In the second part of the proof we then proceed almost word by word as before: we just use the new partite Lemma 14.14 and the new amalgamation technique instead of the ones from the graph case.

Proof of Theorem 14.12. Let $\mathcal{E} \subseteq \mathcal{H}^k(t)$ and $\mathcal{F} \subseteq \mathcal{H}^k(m)$. Choose a positive integer n such that $n \geq GR(|A|, k, m, r)$, where $GR(\cdot)$ is as defined by the Graham-Rothschild partition theorem for parameter sets (Theorem 5.1).

We first construct a suitable hypergraph $\mathcal{F}_0$ satisfying certain coloring properties. Let $(f_i)_{i<z}$ be an enumeration of $[A]\binom{n}{m}$. Furthermore, let a and b be any two distinct elements of A and let for $i < z$ the z-tuple $y_i \in [A]\binom{z}{0}$ be defined by

$$y_i = (\underbrace{a, \ldots, a}_{i \text{ times}}, b, \underbrace{a, \ldots, a}_{(z-1-i) \text{ times}}).$$

Consider the m-parameter word $f_i^* = f_i \times y_i \in [A]\binom{n+z}{m}$. Each f_i^* describes an m-subspace of $[A]\binom{n+z}{0}$. Moreover, $f_i^* \cdot [A]\binom{m}{0} \cap f_j^* \cdot [A]\binom{m}{0} = \emptyset$ for $i < j < z$, i.e., all these subspaces are mutually disjoint. Hence we can define a hypergraph $\mathcal{F}_0 \subseteq \mathcal{H}^k(n+z)$ such that each f_i^* is an embedding of $\mathcal{F}$, viz., let

$$e_{f_j^* \cdot g} \in \mathcal{F}_0 \quad \text{iff} \quad e_g \in \mathcal{F},$$

for all $j < z, i \leq k$ and $g \in [A]\binom{m}{i}$. Observe that each f_i^* induces a crossing subgraph of $\mathcal{F}_0$ isomorphic to $\mathcal{F}$, with one vertex in each partition, and $\mathcal{F}_0$ itself is n-partite. Of course, if we can find a monochromatic $\mathcal{F}_0$-subgraph, then it clearly implies the existence of a monochromatic $\mathcal{F}$-subgraph. However, the trick lies in the following much weaker coloring requirement:

($\star$) For any $h, h' \in [A]\binom{\mathcal{F}_0}{\mathcal{E}}_{part}$ such that $h{\restriction}n = h'{\restriction}n$, we have $\Delta(\mathcal{F}_0[h]) = \Delta(\mathcal{F}_0[h'])$.

In other words, instead of requiring that $\mathcal{F}_0$ is monochromatic, we require that any two partite $\mathcal{E}$-subgraphs of $\mathcal{F}_0$ spanned by the same parts have the same color.

To see that this suffices, consider a coloring $\Delta : \binom{\mathcal{G}}{\mathcal{E}} \to r$ and assume that an $\mathcal{F}_0$-subgraph $\tilde{\mathcal{F}}_0$ of $\mathcal{G}$ satisfying property ($\star$) is given. Then this induces a coloring $\Delta' : [A]\binom{n}{t} \to r$ given by

$$\Delta'(h') = \begin{cases} \Delta(\tilde{\mathcal{F}}_0[h]) & \text{if there exists } h \in \binom{\tilde{\mathcal{F}}_0}{\mathcal{E}} \text{ such that } h{\restriction}n = h' \\ 0 & \text{otherwise.} \end{cases}$$

Note that property ($\star$) implies that Δ' is well-defined. Then by the Graham-Rothschild theorem and choice of n, there exists $f \in [A]\binom{n}{m}$ such that $\Delta'{\restriction}\{f \cdot x \mid x \in [A]\binom{m}{t}\}$ is a constant coloring. As we enumerated $[A]\binom{n}{m}$ we know that $f = f_i$ for some $i < z$. But then $\Delta{\restriction}\binom{\tilde{\mathcal{F}}_0[f_i^*]}{\mathcal{E}}_{part}$ is also a monochromatic coloring, and by the construction $\tilde{\mathcal{F}}_0[f_i^*]$ is isomorphic to $\mathcal{F}$. As each vertex of $\tilde{\mathcal{F}}_0[f_i^*]$ belongs to a distinct partition, we have that $\binom{\tilde{\mathcal{F}}_0[f_i^*]}{\mathcal{E}}_{part}$ coincides with $\binom{\tilde{\mathcal{F}}_0[f_i^*]}{\mathcal{E}}$ and thus we have found a monochromatic $\mathcal{F}$-subgraph.

Next, we construct an n-partite hypergraph $\mathcal{G} \subseteq \mathcal{H}^k(n+n')$ such that for every coloring $\Delta : \binom{\mathcal{G}}{\mathcal{E}}_{part} \to r$ there exists an $\mathcal{F}_0$-subgraph with property ($\star$).

Let $(h_i)_{i<q}$ be an enumeration of $[A]\binom{n}{t}$. According to the partite lemma (Lemma 14.14), let $\mathcal{F}_0^*$ be a t-partite hypergraph satisfying

$$\mathcal{F}_0^* \xrightarrow{\text{part}} ((\mathcal{F}_0)_{h_0})_r^{\mathcal{E}},$$

where $\mathcal{E}$ is viewed as a t-partite graph. Now let $\mathcal{F}_1 = (\mathcal{F}_0)_{*h_0}(\mathcal{F}_0^*)$.

Observe that $\mathcal{F}_1$ has the following property. For any coloring $\Delta : \binom{\mathcal{F}_1}{\mathcal{E}}_{part} \to r$ there exists $f \in \binom{\mathcal{F}_1}{\mathcal{F}_0}_{part}$ which satisfies that for any $h, h' \in \binom{\mathcal{F}_1[f]}{\mathcal{E}}_{part}$ such that $h{\restriction}n = h'{\restriction}n = h_0$, we have $\Delta(\mathcal{F}_1[f \cdot h]) = \Delta(\mathcal{F}_1[f \cdot h'])$. Therefore, we have an $\mathcal{F}_0$-subgraph which satisfies property $(\star)$ when restricted to the $\mathcal{E}$-subgraphs spanned by the partition given by h_0.

Let us assume that we have constructed a hypergraph $\mathcal{F}_i \subseteq \mathcal{H}^k(n + n_i)$ with the similar property as for $\mathcal{F}_1$: for any coloring $\Delta : \binom{\mathcal{F}_i}{\mathcal{E}}_{part} \to r$ there exists $f \in \binom{\mathcal{F}_i}{\mathcal{F}_0}_{part}$ which satisfies that for any $h, h' \in \binom{\mathcal{F}_i[f]}{\mathcal{E}}_{part}$ such that $h{\restriction}n = h'{\restriction}n = h_j$ for some $j < i$, we have $\Delta(\mathcal{F}_i[f \cdot h]) = \Delta(\mathcal{F}_i[f \cdot h'])$. Then, again by the partite lemma (Lemma 14.14), let $\mathcal{F}_i^*$ be a t-partite hypergraph satisfying

$$\mathcal{F}_i^* \xrightarrow{part} ((\mathcal{F}_i)_{h_i})_r^{\mathcal{E}},$$

and let $\mathcal{F}_{i+1} = (\mathcal{F}_i)_{*h_i}(\mathcal{F}_i^*)$. A moment of thought now reveals that $\mathcal{F}_{i+1}$ always contains an $\mathcal{F}_i$-subgraph which is monochromatic with respect to $\mathcal{E}$-copies spanned by partitions given by h_i. But now this $\mathcal{F}_i$ copy further contains an $\mathcal{F}_0$-subgraph for which property $(\star)$ holds for all $\mathcal{E}$-copies spanned by partitions given by $h_0, \ldots, h_{i-1}$ and, by previous observation, also h_i.

Inductively repeating the same argument, we get that $\mathcal{F}_q$ always contains an $\mathcal{F}_0$-subgraph which satisfies property $(\star)$. By the previous observations, this implies the existence of a monochromatic $\mathcal{F}$-subgraph, which finishes the proof. $\qquad\square$

Chapter 15
Ramsey Statements for Random Graphs

Ramsey's theorem implies that for all graphs F and r we have $K_n \to (F)^e_r$, for n large enough. At first sight it is not immediately clear whether this follows from the density of K_n or its rich structure. As it turns out, studying Ramsey properties of random graphs shows that the later is the case, as random graphs give examples of sparse graphs with the desired Ramsey property.

Apparently, Erdős was first to ask whether there exists a graph G such that $G \to (K_3)^e_2$ and, additionally G has small clique size $cl(G)$. Recall that $cl(G)$ denotes the maximal size of a complete subgraph in G. Eventually Folkman (1970) constructed a graph G with $G \to (K_3)^e_2$ and $cl(G) = 3$. Such results are so-called *restricted* Ramsey theorems, as they put up restrictions on the host graph G. E.g., to get a monochromatic triangle in a graph no K_4 should be necessary. So one asks to what extent certain restrictions concerning the appearance of certain substructures which are valid for F can also be satisfied for G, but still $G \to (F)^H_r$ resp. $G \xrightarrow{\text{ind}} (F)^H_r$.

In this chapter we attack this question from a random view point. We consider the Erdős-Rényi random graph $G_{n,p}$ defined as follows: we start with the complete graph K_n and decide for every edge independently whether we keep it (with probability p) or whether it is deleted (with probability $1 - p$). In general, we allow the edge probability p to be a function of n. For a given graph F we then ask the following question: for which edge probabilities $p = p(n, F)$ do we have

$$\lim_{n \to \infty} \Pr[G_{n,p} \to (F)^e_r] = 1.$$

Clearly, if we have $\Pr[G_{n,p} \to (F)^e_r] > 1/2$, say, and we know in addition that $\Pr[G_{n,p} \text{ satisfies property } \mathcal{P}] > 1/2$, then there *exists* a graph G that satisfies property $\mathcal{P}$ *and* has the property that $G \to (F)^e_r$. This idea will allow us in particular to derive the following theorem:

Theorem 15.1. *For every $\ell \geq 3$ and every positive integer r there exists a graph G such that*

H.J. Prömel, *Ramsey Theory for Discrete Structures*,
DOI 10.1007/978-3-319-01315-2_15,
© Springer International Publishing Switzerland 2013

$$cl(G) = \ell \qquad and \qquad G \to (K_\ell)_r^e.$$

We note that Nešetřil and Rödl (1976b) proved a stronger statement. Namely they show that for every graph F and every $r \geq 2$ there exists a graph G with $cl(G) = cl(F)$ such that $G \overset{ind}{\to} (F)_r^e$. For the proof of their result Nešetřil and Rödl use the amalgamation technique that was introduced in the previous section to obtain the desired graph G constructively.

15.1 Rödl-Ruciński's Theorem

The study of random Ramsey theory was initiated by Łuczak et al. (1992). Thereupon, in a series of papers Rödl and Ruciński (1993, 1994, 1995) determined the threshold of $G_{n,p} \to (F)_r^e$, in full generality. Formally, their result reads as follows.

Notation. For a graph G on at least three vertices we set $d_2(G) = (e_G - 1)/ (v_G - 2)$. By $m_2(G)$ we denote for every graph G the so-called maximum 2-density, defined as

$$m_2(G) = \max_{J \subseteq G, v_J \geq 3} d_2(J).$$

If $m_2(G) = d_2(G)$ we say that a graph G is 2-*balanced*, and if in addition $m_2(G) > d_2(J)$ for every subset $J \subset G$ with $v_J \geq 3$, we say that G is strictly 2-*balanced*.

Theorem 15.2 (Rödl, Ruciński). *Let $r \geq 2$ and F be a fixed graph that is not a forest of stars or, in the case $r = 2$, paths of length 3. Then there exist positive constants $c = c(F, r)$, and $C = C(F, r)$ such that*

$$\lim_{n \to \infty} \Pr[G_{n,p} \to (F)_r^e] = \begin{cases} 0 & \text{if } p \leq cn^{-1/m_2(F)} \\ 1 & \text{if } p \geq Cn^{-1/m_2(F)}. \end{cases}$$

For the exceptional case of a star with k edges it is easily seen that the threshold is determined by the appearance of a star with $r(k - 1) + 1$ edges. For paths P_3 of length 3 the 0-statement only holds for $p \ll n^{-1/m_2(P_3)} = n^{-1}$ since, for example, a C_5 with a pending edge at every vertex has density one but cannot be edge-colored with two colors without a monochromatic P_3.

Note that $p = n^{-1/m_2(F)}$ is the density where we expect that every edge is contained in roughly a constant number of copies of F. This observation can be used to provide an intuitive understanding of the bounds of Theorem 15.2. If c is very small, then the number of copies of F is asymptotically almost surely (i.e., with probability $1 - o(1)$ if n tends to infinity) small enough that they are so scattered that a coloring without a monochromatic copy of F can be found. If, on the other hand, C is big then these copies a.a.s. overlap so heavily that every coloring has to induce at least one monochromatic copy of F.

In the remainder of this section we give a proof of Theorem 15.2; our proof is taken from Nenadov and Steger (to appear).

15.2 Proof of the 1-Statement

The proof of the 1-statement requires two tools. The first one is a quantitative strengthening of Ramsey's theorem.

Theorem 15.3. *For every graph F and every constant $r \geq 2$ there exist constants $\alpha > 0$ and n_0 such that for all $n \geq n_0$ every r-coloring of the edges of K_n contains at least αn^{v_F} monochromatic copies of F.*

Proof. From Ramsey's theorem we know that there exists $N := N(F, r)$ such that every r-coloring of the edges of K_N contains a monochromatic copy of F. Thus, in any r-coloring of K_n every N-subset of the vertices contains at least one monochromatic copy of F. As every copy of F is contained in at most $\binom{n - v_F}{N - v_F}$ many N-subsets, the theorem follows e.g. with $\alpha = 1/N^{v_F}$. $\square$

Corollary 15.4. *For every graph F and every $r \in \mathbb{N}$ there exist constants n_0 and $\delta, \epsilon > 0$ such that the following is true for all $n \geq n_0$. For any $E_0, \ldots, E_{r-1} \subseteq E(K_n)$ such that for all $0 \leq i < r$ the set E_i contains at most $\epsilon n^{v(F)}$ copies of F, we have*

$$|E(K_n) \setminus (E_0 \cup \ldots \cup E_{r-1})| \geq \delta n^2.$$

Proof. Let α and n_0 be as given by Theorem 15.3 for F and $r+1$, and set $\epsilon = \alpha/2r$. In addition, let $E_r := E(K_n) \setminus (E_0 \cup \ldots \cup E_{r-1})$, and consider the coloring $\Delta : E(K_n) \rightarrow r + 1$ given by $\Delta(e) = \min\{i \in r + 1 : e \in E_i\}$. By Theorem 15.3 there exist at least αn^{v_F} monochromatic copies of F under coloring Δ, of which, by assumption on the sets E_i, at least $\frac{1}{2}\alpha \cdot n^{v_F}$ must be contained in E_r. As every edge is contained in at most $2e_F \cdot n^{v_F - 2}$ copies of F the claim of the corollary follows e.g. for $\delta = \frac{\alpha}{4e_F}$. $\square$

The second tool that we need is a consequence of the so-called container theorems of Balogh et al. (2012) and Saxton and Thomason (2012). The following theorem is from Saxton and Thomason, who obtained it for all graphs F. Balogh, Morris and Samotij proved a similar statement for all 2-balanced graphs F.

Definition 15.5. *For a given set E and constants $k \in \mathbb{N}$, $s > 0$, let $\mathcal{T}_{k,s}(E)$ be the family of k-tuples of subsets defined as follows,*

$$\mathcal{T}_{k,s}(E) = \{(E_0, \ldots, E_{k-1}) \mid E_i \subseteq E \text{ for } i < k \text{ and } |\bigcup_{i<k} E_i| \leq s\}.$$

Theorem 15.6. *For any graph F and $\epsilon > 0$, there exist $n_0 \in \mathbb{N}$ and $c > 0$ such that the following is true. For every $n \geq n_0$ and $\tau > cn^{-1/m_2(F)}$ there exist $t = t(n)$, pairwise distinct tuples $T_0, \ldots, T_{t-1} \in \mathcal{T}_{c,c\cdot\tau n^2}(E(K_n))$ and sets $C_0, \ldots, C_{t-1} \subseteq E(K_n)$, such that*

(1) *Each C_i contains at most ϵn^{v_F} copies of F,*
(2) *For every graph G on n vertices containing at most $\epsilon \cdot \tau^{e_F} n^{v_F}$ copies of F, there exists $i < t$ such that $T_i \subseteq E(G) \subseteq C_i$. (Here $T_i \subseteq E(G)$ means that all sets contained in T_i are subsets of $E(G)$.)*

With these two tools at hand the proof of the 1-statement of Theorem 15.2 is now easily completed. Actually, they will allow us to prove the following stronger statement:

Theorem 15.7. *Let $r \geq 2$ and F be a fixed graph. Then there exist constants $C = C(F, r)$ and $\gamma = \gamma(F, r) > 0$ such that for all $p \geq Cn^{-1/m_2(F)}$ the random graph $G_{n,p}$ has with probability $1 - o(1)$ the property that for every r-coloring of edges of $G_{n,p}$ there exist at least $\gamma \cdot n^{v_F} p^{e_F}$ monochromatic F-subgraphs in color i for some $0 \leq i < r$.*

Proof. Let ϵ and δ be as provided by Corollary 15.4. Furthermore, let n_0 and $c > 0$ be as provided by Theorem 15.6 with respect to F and ϵ. Denote by $\alpha > 0$ some (small) constant that we will fix below (depending on c, δ and r) and choose C such that $C > c/\alpha$. Let $\mathcal{P}$ denote the property that for all colorings $\Delta : E(G_{n,p}) \to r$ we have at least $\epsilon \alpha^{e_F} p^{e_F} n^{v_F}$ monochromatic F-subgraphs in color i for some $i < r$. We need to show that for $p \geq Cn^{-1/m_2(F)}$ we have

$$\Pr[G_{n,p} \text{ does not satisfy } \mathcal{P}] = o(1).$$

Clearly, this proves the theorem setting $\gamma = \epsilon \alpha^{e_F}$.

Assume that $n \geq n_0$ and that $G_{n,p}$ does not satisfy property $\mathcal{P}$. Then there exists a coloring $\Delta : E(G_{n,p}) \to r$ so that for all $j < r$ the sets $E_j := \Delta^{-1}(j)$ contain at most $\epsilon \alpha^{e_F} p^{e_F} n^{v_F}$ copies of F. By Theorem 15.6 (applied with $\tau := \alpha p$) we have that for every such E_j there exists $1 \leq i_j \leq t(n)$ such that $T_{i_j} \subseteq E_j \subseteq C_{i_j}$ and C_{i_j} contains at most $\epsilon n^{v(F)}$ copies of F. The trivial, but nonetheless crucial observation is that $G_{n,p}$ completely avoids $E(K_n) \setminus (C_{i_0} \cup \ldots \cup C_{i_{r-1}})$, which by Corollary 15.4 has size at least δn^2.

Therefore we can bound the probability that $G_{n,p}$ does not satisfy $\mathcal{P}$ by the probability that there exist tuples $T_{i_0}, \ldots, T_{i_{r-1}}$ that are contained in $G_{n,p}$ such that $E_0(T_{i_0}, \ldots, T_{i_{r-1}}) := E(K_n) \setminus (C_{i_0} \cup \ldots \cup C_{i_{r-1}})$ is edge-disjoint from $G_{n,p}$. Thus

$$\Pr[G_{n,p} \text{ does not satisfy } \mathcal{P}]$$

$$\leq \sum_{i_0,\ldots,i_{r-1}} \Pr[T_{i_0}, \ldots, T_{i_{r-1}} \subseteq G_{n,p} \wedge G_{n,p} \cap E_0(T_{i_0}, \ldots, T_{i_{r-1}}) = \emptyset],$$

where $i_0, \ldots, i_{r-1}$ run over the choices given by Theorem 15.6. Note that the two events in the above probability are independent and can thus be bounded by $p^{|\bigcup_{j<r} T_{ij}^+|} \cdot (1-p)^{\delta n^2}$, where by T_{ij}^+ we denote the union of the sets of the k-tuple T_{ij}. Note that $(1-p)^{\delta n^2} \leq e^{-\delta p n^2}$. The sum can be bounded by first deciding on $s := |\bigcup_{j<r} T_{ij}^+| \leq r \cdot c\tau n^2$, then choosing that many edges ($\binom{\binom{n}{2}}{s}$ choices) and finally deciding for every edge in which sets of the tuples T_{ij} it appears $((2^{rc})^s$ choices). Together, this gives

$$\Pr[G_{n,p} \text{ does not satisfy } \mathcal{P}] \leq e^{-\delta n^2 p} \cdot \sum_{s=0}^{r \cdot c\tau n^2} \binom{\binom{n}{2}}{s}(2^{rc})^s p^s$$

$$\leq e^{-\delta n^2 p} \cdot \sum_{s=0}^{r \cdot c\tau n^2} \left(\frac{e2^{rc} n^2 p}{2s} \right)^s.$$

Recall that $\tau = \alpha p$. By choosing α sufficiently small (with respect to c, δ and r) we may assume that

$$\sum_{s=0}^{r \cdot c\tau n^2} \left(\frac{e2^{rc} n^2 p}{2s} \right)^s \leq n^2 \cdot \left(\frac{e2^{rc}}{2rc \cdot \alpha} \right)^{rc\alpha \cdot n^2 p} \leq e^{\frac{1}{2}\delta n^2 p},$$

and thus $\Pr[G_{n,p} \text{ does not satisfy } \mathcal{P}] = o(1)$, as desired. $\square$

With Theorem 15.7 at hand, we can now easily prove the restricted Ramsey theorem for cliques.

Proof of Theorem 15.1. Let $p = Cn^{-1/m_2(K_\ell)} = Cn^{-2/(\ell+1)}$, where C is given by Theorem 15.7. We claim that there exist positive constants α, β and γ such that $G_{n,p}$ satisfies with positive probability the following three properties simultaneously:

(i)   Every edge is contained in at most $\alpha \log n$ copies of K_ℓ.
(ii) The number of copies of $K_{\ell+1}$ is bounded by βn.
(iii) For every r-coloring of the edges we have at least $\gamma n^{2-\frac{2}{\ell+1}}$ monochromatic copies of K_ℓ.

Clearly, every graph that satisfies the above properties simultaneously, can be used to construct the desired graph G. Simply delete an edge from every copy of $K_{\ell+1}$. By (i) and (ii) this will delete at most $\alpha\beta n \log n$ copies of K_ℓ. As every r-coloring contains a lot more monochromatic copies, one of it will still be present.

So it remains to show that all three properties hold simultaneously with positive probability. Clearly, it suffices to show that every property individually holds with probability at least $3/4$. For (iii) this follows immediately from Theorem 15.7. (ii) follows from Markov's inequality, as the expected number of copies of $K_{\ell+1}$ in $G_{n,p}$ is bounded by

$$n^{\ell+1} p^{\binom{\ell+1}{2}} = C^{\binom{\ell+1}{2}} \cdot n.$$

The proof of (i) is a bit more subtle. Let us first look at the case $\ell = 3$. Then the expected number of edges that are contained in $k := \log n$ triangles is bounded by

$$n^2 \cdot \binom{n}{k} \cdot p^{2k+1} \le n^2 \left(\frac{en}{k}\right)^k (Cn^{-1/2})^{2k+1},$$

which is easily seen to be $o(1)$. So (i) again follows from Markov's inequality in this case. For $\ell \ge 4$ we have to be more careful, as copies of K_ℓ that sit on the same edge may overlap in various ways. In this case the claim follows, for example, from Spencer's extension theorem (Spencer 1990). We omit the details. $\square$

15.3   Proof of the 0-Statement

We need to show that with high probability the edges of a random graph $G_{n,p}$ with $p = cn^{-1/m_2(F)}$, with $0 < c = c(F) < 1$ small enough, can be colored in such a way that we have no monochromatic F-subgraph. If $m_2(F) = 1$ we have $p \le cn^{-1}$ with $c < 1$. It is well-known that then every component of $G_{n,p}$ is a.a.s. either a tree or a unicyclic graph (see Erdős and Rényi 1960). One easily checks that we can color each such component without a monochromatic copy of F if F is not a star and not a path of length 3 (or $r \ge 3$ in the latter). In the following we thus assume that $m_2(F) > 1$.

Observe that we may also assume without loss of generality that $r = 2$ and that F its strictly 2-balanced. If not, replace F by a minimal subgraph F' with the same 2-density. Clearly, if we find a 2-coloring of the edges of $G_{n,p}$ without a monochromatic copy of F' this 2-coloring will also contain no monochromatic copy of F.

The expected number of copies of F on any given edge is bounded by

$$2e_F \cdot n^{v_F-2} \cdot p^{e_F-1} = 2e_F \cdot c^{e_F-1}.$$

That is, for $0 < c < 1$ small enough we do not expect more than one copy. Observe that if an edge is contained in at most one copy of F, then this edge can always be colored such that it will not be part of a monochromatic copy of F: just color the edge arbitrarily if the remainder of the copy uses both colors, otherwise use the opposite color. We will now make this idea more formal.

Let e be an edge in $G_{n,p}$. Assume that $G_{n,p} - e$ is 2-colorable without a monochromatic copy of F. Consider any such coloring. If this coloring cannot be extended to e then there has to exist both a red and a blue copy of $F - \hat{e}$ such that e completes both of these copies to a copy of F. Clearly, the blue and the red copy of $F - \hat{e}$ are edge disjoint. We thus conclude that there exist at least two copies of

F which intersect only in e. In other words: a necessary obstruction for extending a coloring from $G - e$ to G is that e is contained in at least two copies of F that only intersect in e.

To formalize this idea, call an edge e *closed* in G if it is contained in at least two copies of F whose edge sets intersect exactly in e. (Note: we do allow that the vertex sets of these copies intersect in more than two vertices.) Otherwise we call the edge *open*. With this notion at hand we can now formulate the following algorithm for obtaining the desired 2-coloring of $G_{n,p}$:

> $\hat{G} := G_{n,p}$;
> **while** there exists an open edge e in $\hat{G}$ **do**
> $\qquad \hat{G} \leftarrow \hat{G} - e$;
> color $\hat{G}$;
> add the edges in reverse order and color them appropriately.

The critical point, of course, is the statement 'color $\hat{G}$'. We need to show that this step is indeed possible.

Observe that after termination of the **while**-loop the graph $\hat{G}$ has the following property: every edge of $\hat{G}$ is closed. It is easy to see that $\hat{G}$ is actually the (unique) maximal subgraph of $G_{n,p}$ with the property that every edge is closed (within this subgraph). We call $\hat{G}$ the F-core of $G_{n,p}$.

We now further refine $\hat{G}$. Consider an auxiliary graph G_F defined as follows: the set of vertices correspond to the set of copies of F in $\hat{G}$ and two vertices are connected by an edge if and only if the corresponding copies of F have at least one edge in common. Since every edge of $\hat{G}$ belongs to a copy of F, the connected components of G_F naturally partition the edges of $\hat{G}$ into equivalence classes. Observe that, by definition, each equivalence class (an F-*component* for short) can be colored separately in order to find a valid coloring of the F-core. Note also that within $\hat{G}$ the F-components need not necessarily form components. For example, a cube is a C_4-component: every edge of the cube is in two otherwise edge-disjoint C_4's. If we now attach two cubes at a vertex then the two cubes are connected – but for the purpose of obtaining an edge coloring without a monochromatic C_4 we can, of course, still consider both cubes separately.

The core of the proof is the following lemma which states that with high probability every F-component in the F-core of $G_{n,p}$ has constant size.

Lemma 15.8. *Let F be a strictly 2-balanced graph with $e_F \geq 3$. There exist $c = c(F) > 0$ and $L = L(F) > 0$ such that if $p \leq cn^{-1/m_2(F)}$ then w.h.p. every F-component of the F-core of $G_{n,p}$ has size at most L.*

We defer the proof of this lemma to the end of this section and first show how it can be used to complete the proof of the 0-statement. For that we make use of the following result of Rödl and Ruciński (1993) that states that graphs with G with small enough density do not have the Ramsey property.

Lemma 15.9. *Let G and F be two graphs. If $m(G) \leq m_2(F)$ and $m_2(F) > 1$ then $G \nrightarrow (F)_2^e$.*

Proof. We first consider the case that $v_F \geq 4$. Observe that we may assume without loss of generality that F is strictly 2-balanced. The assumption that F is strictly 2-balanced implies that $m_2(F) = \frac{e_F - 1}{v_F - 2} > \frac{e_F - \delta(F) - 1}{v_F - 3}$ from which we deduce $m_2(F) < \delta(F) \leq \delta_{max}(F)$ (cf. p. 130 for the definition of $\delta_{max}(G)$). If $\chi(F) \geq 3$, then $G \nrightarrow (F)_2^e$ follows from Lemma 12.5. Therefore, in the following, we assume that F is bipartite. Then $e_F \leq \frac{1}{4} v_F^2$ implies that

$$m_2(F) \leq m(F) + \frac{1}{2} \quad \text{with equality if and only if } e_F = \frac{1}{4} v_F^2.$$

If $m_2(F) = k + x$ for some $k \in \mathbb{N}$ and $\frac{1}{2} \leq x < 1$ we thus have $m(F) > k$ whenever $x > \frac{1}{2}$ or $e_F < \frac{1}{4} v_F^2$. In this case we have $m(G) \leq k + 1 \leq 2k = 2\lfloor m(F) - \epsilon \rfloor$ and F and Lemma 12.8 concludes the proof of the theorem in this case. So we may assume that $x = \frac{1}{2}$ and $e_F = \frac{1}{4} v_F^2$. Then, $v_F = 2\ell$ for some $\ell \in \mathbb{N}$, and thus $m_2(F) = (\ell^2 - 1)/(2\ell - 2) = \frac{1}{2}(\ell + 1)$. That is, $k = \frac{1}{2}\ell$ and so $ar(F) = e_F/(v_F - 1) > \frac{1}{4} v_F = k$. By (12.1) we also have $ar(G) \leq m(G) + \frac{1}{2} \leq m_2(F) + \frac{1}{2} = k + 1$. Thus $ar(G) \leq k + 1 \leq 2k \leq 2\lfloor ar(F) - \epsilon \rfloor$ and F satisfies the property (i) and Lemma 12.8 also concludes the proof of the theorem in this case.

Finally, assume $m_2(F) = k + x$ for some $k \in \mathbb{N}$ and $0 \leq x < 0.5$. Then the fact that for every graph G we have $\delta_{max}(G) \leq m(G) \leq 2m_2(F)$ implies that $\delta_{max}(G) \leq 2k$, as $\delta_{max}(G)$ is integral. On the other hand, we have already shown that the assumption that F is strictly 2-balanced implies that $m_2(F) < \delta(F)$. The fact that $\delta(F)$ is integral thus implies $\delta(F) \geq k + 1$, and Lemma 12.9 concludes the proof of the theorem in this case.

It remains to consider the case $v_F = 3$. Observe that in this case the only graph with $m_2(F) > 1$ is the triangle, i.e., $F = K_3$. Here we proceed similarly as in the proof of Lemma 12.9. I.e., we construct a sequence $v_1, v_2, \ldots,$ by choosing v_i as a vertex of minimum degree in $G - \{v_1, \ldots, v_{i-1}\}$, with the additional condition that the neighborhood of V_i in $G - \{v_1, \ldots, v_{i-1}\}$ is not a K_4. If we do not find a vertex that satisfies this property then we stop. As $\delta_{max}(G) \leq 2m(G) \leq 4$ we will always find a vertex with degree at most 4. Also note that if the minimum degree *is* 4, then the graph is 4-regular. That is, the above process can only stop if *every* vertex has degree 4 *and* has the property that its neighborhood induces a K_4. That is, if we cannot find a vertex v_i, then $G' := G - \{v_1, \ldots, v_{i-1}\}$ is a union of vertex-disjoint K_5's. As K_5 can be 2-colored without inducing a monochromatic triangle we can thus 2-color G' without a monochromatic triangle. Now we proceed again as in the previous proof and color the remaining vertices in reverse order. By construction vertex v_i has degree at most 4 into $G[\{v_i, \ldots, v_{v_H}\}]$ and the neighborhood of v_i in $G[\{v_i, \ldots, v_{v_H}\}]$ is not a K_4. A simple case checking shows that however the neighborhood of v_i is colored without a monochromatic triangle there exists always an extension of the coloring to the edges incident to v_i so that no monochromatic triangle is generated. $\square$

With Lemmas 15.8 and 15.9 at hand, the proof of the 0-statement of Theorem 15.2 is straightforward.

Proof of Theorem 15.2 (0-statement). Let us first consider the case $m_2(F) > 1$. Recall that we may assume w.l.o.g. that F is strictly 2-balanced. Choose $c = c(F)$ and $L = L(F)$ according to Lemma 15.8. Then $G_{n,p}$ a.a.s. has the property that every F-component of the F-core of $G_{n,p}$ has size at most L.

Observe that there exist only constantly many different graphs on L vertices. Let H be one such graph and choose $H' \subseteq H$ such that $m(H) = e_{H'}/v_{H'}$. Then the expected number of copies of H' in $G_{n,p}$ is bounded by $n^{v_{H'}} p^{e_{H'}}$. Observe that for $p = cn^{-1/m_2(F)}$ we have $n^{v_{H'}} p^{e_{H'}} = o(1)$ whenever $m(H) = e_{H'}/v_{H'} > m_2(F)$. It thus follows from Markov's inequality that for $p \le cn^{-1/m_2(F)}$ a.a.s. there is no copy of H', and hence no copy of H in $G_{n,p}$. That is, a.a.s. every subgraph G of $G_{n,p}$ of size $|V(G)| \le L$ satisfies $m(G) \le m_2(F)$.

Combining both properties we thus get: a.a.s. all F-components of the F-core of $G_{n,p}$ satisfy $m(G) \le m_2(F)$ and Theorem 15.9 thus implies that there exists a 2-edge-coloring for G without a monochromatic copy of F. The union of these edge colorings of all F-components thus yields the desired coloring of the F-core of $G_{n,p}$. As explained above this coloring can be extended to a valid coloring of $G_{n,p}$.   □

In the remainder of this section we prove Lemma 15.8. We start by collecting some properties of strictly 2-balanced graphs.

Lemma 15.10. *If F is strictly 2-balanced, then F is 2-connected.*

Proof. Clearly, F is connected. As then $(e_F - 2)/(v_F - 3) \ge (e_F - 1)/(v_F - 2)$, we deduce that F cannot contain a vertex of degree 1. Assume there exists $v \in V(F)$ that is a cut vertex. Then there exist subgraphs F_1 and F_2 that both contain at least three vertices such that $F_1 \cup F_2 = F$ and $V(F_1) \cap V(F_2) = \{v\}$. As F is strictly 2-balanced we get

$$e_F - 2 = (e_{F_1} - 1) + (e_{F_2} - 1) < m_2(F) \cdot (v_{F_1} - 2 + v_{F_2} - 2) = m_2(F) \cdot (v_F - 3).$$

(Here we used that $a/b < x$ and $c/d < x$ implies $(a + c)/(b + d) < x$.) As $m_2(F) = (e_F - 1)/(v_F - 2)$ (as F is balanced), this implies $e_F < v_F - 1$. A contradiction.   □

Lemma 15.11. *Let F be strictly 2-balanced and let G be an arbitrary graph. Construct a graph $\hat{G}$ by attaching F to an edge e of G. Then $\hat{G}$ has the property that if $\hat{F}$ is a copy of F in $\hat{G}$ that contains a least one vertex from $F - e$, then $\hat{F} = F$.*

Proof. Assuming the opposite, let $\hat{F}$ be a copy of F which violates the claim. Set $F_g = \hat{F}[V(G)]$ and $F_f = \hat{F}[V(F)]$ and, if $e \notin F_f$, add the edge e to F_f. Then F_g and F_f are strict subgraphs of F, thus

$$\frac{e_{F_g} - 1}{v_{F_g} - 2} < m_2(F) \quad \text{and} \quad \frac{e_{F_f} - 1}{v_{F_f} - 2} < m_2(F)$$

since F is strictly 2-balanced. Furthermore, since every strictly 2-balanced graph is, by Lemma 15.10, 2-connected, it follows that both vertices of e belong to $V(\hat{F})$, thus $v_F = v_{F_g} + v_{F_f} - 2$. Finally, it is easy to see that $e_F = e_{F_g} + e_{F_f} - 1$ regardless of whether $e \in \hat{F}[V(F)]$ or not. This, however, yields a contradiction, as

$$m_2(F) = \frac{e_F - 1}{v_F - 2} = \frac{e_{F_g} - 1 + e_{F_f} - 1}{v_{F_g} - 2 + v_{F_f} - 2} < m_2(F).$$

$\square$

In order to prove Lemma 15.8 we define a process that generates F-components iteratively starting from a single copy of F.

Let G' be an F-component of the F-core of $G_{n,p}$. Then G' can be generated by starting with an arbitrary copy of F in G' and repeatedly attaching copies of F to the graph constructed so far.

> Let F_0 be a copy of F in G',
> $\ell \leftarrow 0; \hat{G} \leftarrow F_0$;
> **while** $\hat{G} \neq G'$ **do**
>> $\ell \leftarrow \ell + 1$;
>> **if** $\hat{G}$ contains an open edge **then**
>>> let $\ell' < \ell$ be the smallest index such that
>>>> $F_{\ell'}$ contains an open edge;
>>> let e be any open edge in $F_{\ell'}$;
>>> let F_ℓ be a copy of F in G' that contains e but is
>>>> not contained in $\hat{G}$;
>> **else**
>>> let F_ℓ be a copy of F in G' that is not contained
>>>> in $\hat{G}$ and intersects $\hat{G}$ in at least one edge;
>> $\hat{G} \leftarrow \hat{G} \cup F_\ell$;

We will eventually prove Lemma 15.8 by a first moment argument. More precisely, we consider all sequences $(F_0, F_1, \ldots,)$ that generate F-components and multiply the number of choices for such a sequence with the probability that the sequence is contained in $G_{n,p}$. In order to be able to bound this number more precisely we first collect some properties of this process. Consider a copy F_ℓ for $\ell \geq 1$. We distinguish two cases: (a) F_ℓ intersects $\hat{G} := \bigcup_{i < \ell} F_i$ in *exactly* two vertices (that, by definition of the algorithm, have to form an edge), i.e. F_ℓ intersects $\hat{G}$ in exactly one edge (we call this a regular copy) and (b) F_ℓ intersects $\hat{G}$ in some subgraph J with $v_J \geq 3$ (we call this a degenerate copy).

For $0 \leq i \leq \ell$ we say that the copy F_i is *fully-open* at time ℓ if F_i is a regular copy (or $i = 0$) and no vertex of $V(F_i) \setminus (\bigcup_{i' < i} V(F_{i'}))$, is touched by any of the copies $F_{i+1}, \ldots, F_\ell$. Note that F_0 is fully-open only at time 0. Also note that, by Lemma 15.11, every fully-open copy at time $\ell \geq 1$ contains exactly $e_F - 1$ open edges.

For the analysis of the algorithm it is important to keep track of fully-open components. For doing so we introduce the following definition. For $\ell \geq 1$ let

$$\kappa(\ell) = |\{0 \leq i < \ell \mid F_i \text{ fully-open at time } \ell - 1 \text{ but not at time } \ell\}|.$$

Clearly, a regular copy can 'destroy' at most one fully-open copy (as it intersects $\hat{G}$ in exactly one edge). Thus $\kappa(\ell) \leq 1$ if F_ℓ is a regular copy. A degenerate copy on the other hand intersects one F_i in an edge and may destroy up to $v_F - 2$ additional regular copies. Thus, $\kappa(\ell) \leq v_F - 1$ if F_ℓ is a degenerate copy.

Lemma 15.12. *For any sequence $F_i, \ldots, F_{i+e_F-2}$ of consecutive regular copies such that $\kappa(i) = 1$ we have $\kappa(i + 1) = \ldots = \kappa(i + e_F - 2) = 0$.*

Proof. As F_i is a regular copy we know that F_i intersects some copy $F_{i'}$, $i' < i$, in exactly one edge. As $\kappa(i) = 1$ we know that $F_{i'}$ was fully-open at time $i - 1$. Thus at time $i - 1$ the copy $F_{i'}$ had $e_F - 1$ open edges (resp. e_F, if $i' = 0$) and the intersection of F_i with $F_{i'}$ is one of these open edges. At time $i + 1$ the copy $F_{i'}$ thus still has at least $e_F - 2$ open edges and since it was chosen by the process at step i, it will be chosen again in every consecutive step as long as it has an open edge. It easily follows from Lemma 15.11 that every regular copy closes at most one open edge, thus each of the copies $F_{i+1}, \ldots, F_{i+e_F-2}$ intersects $F_{i'}$ in exactly one open edge, which implies $\kappa(i + 1) = \ldots = \kappa(i + e_F - 2) = 0$. □

Next we estimate the number of fully-open copies at time ℓ as a function of the number of regular and degenerate copies. Let us denote with $\mathrm{reg}(\ell)$ and $\deg(\ell)$ the number of copies F_i, $1 \leq i \leq \ell$, which are regular, resp. degenerate. Furthermore, we denote with $f_o(\ell)$ the number of fully-open copies at time ℓ.

Lemma 15.13. *For every $\ell \geq 1$, assuming the process doesn't stop before adding the ℓ-th copy, we have*

$$f_o(\ell) \geq \mathrm{reg}(\ell)(1 - 1/(e_F - 1)) - \deg(\ell) \cdot v_F.$$

Proof. Let $\phi(\ell) := f_o(\ell) - \mathrm{reg}(\ell)(1 - 1/(e_F - 1)) + \deg(\ell) \cdot v_F$. We need to show that $\phi(\ell) \geq 0$ for all $\ell \geq 1$. We actually prove something slightly stronger, namely:

$$\phi(\ell) \geq \begin{cases} 1, & \text{if } F_\ell \text{ is a degenerate copy,} \\ 0, & \text{otherwise.} \end{cases}$$

We prove this by induction on ℓ. One easily checks that the claim holds for $\ell = 1$: if F_1 is a regular copy then $\phi(1) = 1/(e_F - 1)$ and otherwise $\phi(1) = v_F$. Consider some $\ell \geq 2$. If F_ℓ is a degenerate copy then $\phi(\ell) - \phi(\ell - 1) = v_F - \kappa(\ell) \geq 1$ (recall that $\kappa(\ell) \leq v_F - 1$ if F_ℓ is a degenerate copy) and the claim follows. If F_ℓ is a regular copy let $\ell' := \max\{1 \leq \ell' < \ell \mid \kappa(\ell') > 0 \text{ or } F_{\ell'}\text{is a degenerate copy}\}$. (Note that ℓ' is well defined, as $\kappa(1) = 1$.) Then

$$\phi(\ell) - \phi(\ell') = (\ell - \ell')/(e_F - 1) - \kappa(\ell).$$

If $F_{\ell'}$ is a degenerate copy, the claim follows from $\phi(\ell') \geq 1$ (recall that $\kappa(\ell) \leq 1$ if F_ℓ is a regular copy). Otherwise, $\kappa(\ell') > 0$ by the definition of ℓ' and thus $\kappa(\ell') = 1$ since $F_{\ell'}$ is a regular copy. If also $\kappa(\ell) = 1$, then Lemma 15.12 implies $\ell \geq \ell' + (e_F - 1)$, and thus $\phi(\ell) \geq 0$ also in this case. Finally, if $\kappa(\ell) = 0$ then we trivially have $\phi(\ell) \geq 0$. $\qquad\square$

If $f_0(\ell) > 0$ for some $\ell \geq 1$, then F_ℓ cannot be the last copy in the process because there exists at least one edge which is still open. Furthermore, from Lemma 15.13 we have that after adding L copies, out of which at most ξ were degenerate, there are still at least

$$(L - \xi)(1 - 1/(e_F - 1)) - \xi \cdot v_F \tag{15.1}$$

fully-open copies at time L.

Proof of Lemma 15.8. As said above, our goal is to complete the proof by a first moment argument over all sequences $(F_0, F_1, \dots,)$ that generate F-components. In a first moment calculation we have to multiply the number of choices for a graph F_ℓ with the probability that the chosen copy of F is in $G_{n,p}$. For a regular copy where F_ℓ is attached to an open edge, we get that this term is bounded by

$$2e_F^2 \cdot n^{v_F - 2} \cdot p^{e_F - 1} \leq 2e_F^2 \cdot c^{e_F - 1} < \tfrac{1}{2}, \tag{15.2}$$

for $0 < c < 1/(4e_F^2)$. Here the term $2e_F^2$ bounds the number of choices of the open edge in $F_{\ell'}$ (at most e_F choices) times the number of choices for the edge in F_ℓ that is merged with this open edges (e_F choices) times 2 for the orientation. For a regular copy F_ℓ that is attached to a closed edge we have to replace the first factor e_F by, say, $\ell \cdot e_F$, as the edge e to which the new copy F_ℓ is attached can be any of the previously added edges.

To bound the term for degenerate copies, observe first that for every subgraph $J \subsetneq F$ with $v_J \geq 3$ we have

$$\frac{e_F - 1}{v_F - 2} = m_2(F) > \frac{e_J - 1}{v_J - 2} \quad \text{and thus} \quad \frac{e_F - e_J}{v_F - v_J} = \frac{(e_F - 1) - (e_J - 1)}{(v_F - 2) - (v_J - 2)} > m_2(F).$$

We may thus choose an $\alpha > 0$ so that

$$(v_F - v_J) - \frac{e_F - e_J}{m_2(F)} < -\alpha \quad \text{for all } J \subsetneq F \text{ with } v_J \geq 3.$$

We can now bound the case that the copy F_ℓ is a degenerate copy by

$$\sum_{J \subsetneq F, v_J \geq 3} (\ell \cdot v_F)^{v_J} \cdot n^{v_F - v_J} \cdot p^{e_F - e_J} < (\ell \cdot v_F \cdot 2^{e_F})^{v_F} \cdot n^{-\alpha}, \tag{15.3}$$

with room to spare.

We now do a union bound. For that we choose ξ such that $\xi \cdot \alpha > v_F + 1$ and then choose L such that the term in (15.1) is positive. (Observe that L is a constant that only depends on the forbidden graph F.) Finally, choose $\ell_0 = (v_F + 1) \log_2 n + \xi$.

Consider first all sequences with the property that $F_{\ell'}$, for $\ell' \leq \ell_0$, is the ξth degenerate copy. Then the expected number of subgraphs in $G_{n,p}$ that can be built by such a sequence is at most

$$\sum_{\ell' \leq \ell_0} \binom{\ell'-1}{\xi-1} n^{v_F} \cdot [(\ell_0 v_F 2^{e_F})^{v_F} \cdot n^{-\alpha}]^{\xi} \cdot L^L \leq n^{v_F} \cdot o(n) \cdot n^{-\alpha \cdot \xi} = o(1),$$

by choice of ξ. Here we used (15.3) and the fact that regular copies contribute a term of at most $1/2 < 1$ (and can thus be ignored) if they occur after step L. Each regular copy before step L on the other hand can contribute at most a factor of L.

So we know that within the first ℓ_0 copies we have less than ξ degenerate ones. Then the choice of L implies that the sequence that generates G' either has length less than L (which is fine) or length at least ℓ_0. It thus suffices to consider all sequences of length ℓ_0. The expected number of subgraphs in $G_{n,p}$ that can be built by such a sequence is at most

$$\sum_{k < \xi} \binom{\ell_0}{k} n^{v_F} \cdot [(\ell_0 v_F 2^{e_F})^{v_F} \cdot n^{-\alpha}]^k \cdot L^L \cdot 2^{-(\ell_0 - k)}$$

$$\leq n^{v_F} \cdot o(n) \cdot n^{-(v_F + 1)} = o(1),$$

by choice of ℓ_0. This concludes the proof of Lemma 15.8 and thus also the proof of the 0-statement. $\square$

Chapter 16
Sparse Ramsey Theorems

Sparse Ramsey theorems for graphs originated with investigations of graphs having large chromatic number and high girth (where the girth of a graph is the length of the smallest cycle in G). Note that this can be viewed as a special kind of restricted graph Ramsey problem. Namely the question whether can we find for every r and ℓ a graph G with girth at least ℓ such that $G \rightarrow (K_2)_r^v$.

Apparently Tutte was the first to look for such graphs. He showed in Descartes (1948) that graphs without triangles can have arbitrary large chromatic number. Later on this result was rediscovered several times (e.g., Mycielski 1955; Zykov 1952). Eventually Erdős gave a complete solution by showing that there exist graphs of arbitrary large chromatic number and, simultaneously, arbitrary high girth (Erdős 1959).

In this section we show how the probabilistic method can be used to establish the existence of sparse Ramsey families for various structures. In Sect. 16.1 we give a first example of the probabilistic method by showing the existence of graphs with simultaneously large girth and large chromatic number. This result from Erdős (1959), resp., rather its proof, is the source of the probabilistic method. In Sect. 16.2 we consider sparse Ramsey families of sets and in Sect. 16.3 sets of integers carrying an arithmetic structure. The final section contains some results for parameter sets.

16.1 Sparse Graphs and Hypergraphs

In this section we give a probabilistic proof for the existence of sparse graphs with large chromatic number.

Theorem 16.1 (Erdős). *Let r and ℓ be positive integers. Then, for n sufficiently large there exists a graph G on n vertices with girth larger than ℓ and chromatic number larger than r.*

H.J. Prömel, *Ramsey Theory for Discrete Structures*,
DOI 10.1007/978-3-319-01315-2_16,
© Springer International Publishing Switzerland 2013

Recall that the girth of a graph $G = (V, E)$ is the length of the shortest cycle in G. Note that $\text{girth}(G) > \ell$ is equivalent to saying that every set of ℓ vertices contains at most $\ell - 1$ edges. For this reason graphs with a large girth are called (locally) sparse. The chromatic number of G is the least positive integer r such that there exists an r-coloring of the vertices of G without any monochromatic edge. Equivalently, the chromatic number of G is larger than r if and only if for every r-coloring of the vertices of G there exists a monochromatic edge. Using the Ramsey arrow this can be expressed as $G \to (K_2)_r^v$.

The following two observations are crucial for the method. The first one shows that for every r-coloring $\Delta : n \to r$ the number of monochromatic pairs has the same order of magnitude as the number of all pairs. The second observation bounds the number of cycles.

Observation 16.2. *There exists $c = c(r) > 0$ so that for every r-coloring $\Delta : n \to r$ we have that*

$$|\{\{a, b\} \in [n]^2 \mid \Delta(a) = \Delta(b)\}| \geq cn^2.$$

Proof. By the pigeonhole principle, for every r-coloring $\Delta : n \to r$ there exists a color that is used at least $\lceil n/r \rceil$ times. This color thus induces at least $\binom{\lceil n/r \rceil}{2}$ monochromatic pairs. $\square$

Observation 16.3. *The complete graph on n vertices contains $\frac{1}{2} \cdot \binom{n}{\ell} \cdot (\ell - 1)! \leq n^\ell$ many cycles of length ℓ.* $\square$

Proof of Theorem 16.1. Let δ be a positive real such that $\delta \cdot \ell < 1$. Let

$$p = n^{-1+\delta}$$

and consider the random graph $G_{n,p}$. Denote by X the number of cycles of length less or equal than ℓ in $G_{n,p}$. By Observation 16.3 we have that

$$\mathbb{E}[X] \leq \sum_{3 \leq \tilde{\ell} \leq \ell} n^{\tilde{\ell}} \cdot p^{\tilde{\ell}} \leq \ell \cdot n^{\ell \cdot \delta}.$$

From Markov's inequality, we deduce that

$$\text{Prob}[X < 2\ell n^{\ell \cdot \delta}] \geq \tfrac{1}{2}. \tag{16.1}$$

Using a union bound argument and Observation 16.2 we conclude that

$$\text{Prob}[\exists \Delta : n \to r \text{ s.t. } |\{\{a, b\} \in E(G_{n,p}) \mid \Delta(a) = \Delta(b)\}| \leq n] \tag{16.2}$$

$$\leq r^n \cdot \binom{\binom{n}{2}}{n} \cdot (1 - p)^{cn^2 - n}$$

$$\leq r^n \cdot n^{2n} \cdot e^{-p \cdot cn^2 + p \cdot n} = o(1),$$

where the last equality follows from the choice of p. Putting (16.1) and (16.2) together shows that, for n sufficiently large,

$$\text{Prob}[\forall \Delta : n \to r : |\{\{a, b\} \in E(G_{n,p}) \mid \Delta(a) = \Delta(b)\}| > n \text{ and } X < 2\ell n^{\ell \cdot \delta}] > 0.$$

From this it follows in particular that there *exists* a graph G on n vertices that satisfies both properties simultaneously, that is,

(1) G contains at most $2\ell n^{\ell \cdot \delta}$ cycles of length less or equal than ℓ, and
(2) For every r-coloring $\Delta : n \to r$ the graph G contains at least n monochromatic edges.

Note that for all n sufficiently large we also have $2\ell n^{\ell \cdot \delta} < n$ (recall that we have chosen $\delta > 0$ such that $\ell \cdot \delta < 1$). Thus, deleting one edge from each cycle of length less or equal than ℓ yields a graph G' with girth larger than ℓ and, by (2), with chromatic number still larger than r. $\qquad\qquad\square$

Using essentially the same method one can establish the existence of sparse m-uniform hypergraphs with large chromatic number. Even more can be shown, namely the existence of sparse and selective m-uniform hypergraphs.

Recall that an hypergraph $H = (n, E)$ is m-uniform if the set of edges satisfies $E \subseteq [n]^m$. Such a hypergraph is called selective (cf. Sect. 12.1) if for every coloring $\Delta : n \to \omega$ there exists an edge $X \in E$ which is either constantly colored or colored one-to-one, i.e., $\Delta \upharpoonright X$ is constant or one-to-one. The following theorem is due to Nešetřil and Rödl (1978b).

Theorem 16.4. *Let m and ℓ be positive integers. Then, for n sufficiently large, there exists a selective m-uniform hypergraph $H = (n, E)$ with girth larger than ℓ.*

First we derive observations analogous to Observations 16.2 and 16.3. For a coloring $\Delta : n \to \omega$ we denote by $\text{sel}(\Delta)$ the set of those $Y \in [n]^m$ such that $\Delta \upharpoonright Y$ is constant or one-to-one, i.e.,

$$\text{sel}(\Delta) = \{Y \in [n]^m \mid \Delta \upharpoonright Y \text{ is constant or } \Delta \upharpoonright Y \text{ is one-to-one}\}.$$

Observation 16.5. *For every m there exists $c = c(m) > 0$ such that the following is true. For every coloring $\Delta : n \to \omega$, where $n \geq n(m)$ is sufficiently large, we have that*

$$|\text{sel}(\Delta)| \geq cn^m.$$

Proof. Let M be such that for every mapping $\Delta : M \to \omega$ there exists a set $Y \in [M]^m$ which is colored constantly or one-to-one. It is easy to show that $M = (m - 1)^2 + 1$ suffices.

For every coloring $\Delta : n \to \omega$, each $X \in [n]^M$ contributes at least one $Y \in [X]^m$ to $sel(\Delta)$. On the other hand, every $Y \in [n]^m$ is contained in $\binom{n-m}{M-m}$ M-element subsets $X \in [n]^M$. Hence, for every coloring $\Delta : n \to \omega$ it follows that

$$|\mathrm{sel}(\Delta)| \geq \binom{n}{M} / \binom{n-m}{M-m} \geq \frac{1}{M^m} \cdot n^m,$$

for n sufficiently large. $\hspace{8cm}$ □

Recall that a cycle of length ℓ in an m-uniform hypergraph $H = (n, E)$ with $E \subseteq [n]^m$ is given by a sequence $x_0, \ldots, x_{\ell-1}$ of mutually distinct vertices and a sequence $X_0, \ldots, X_{\ell-1}$ of mutually distinct edges in E such that $x_i \in X_i \cap X_{i+1}$ for $i < \ell - 1$ and $x_{\ell-1} \in X_{\ell-1} \cap X_0$.

Observation 16.6. *The complete m-uniform hypergraph $H = (n, [n]^m)$ on n vertices contains at most $n^{\ell \cdot (m-1)}$ cycles of length ℓ.*

Proof. We have $\binom{n}{\ell} \leq n^\ell$ possible choices for the vertices $x_0, \ldots, x_{\ell-1}$. Having these at hand, we still have $\binom{n}{m-2} \cdot (\binom{n}{m-2} - 1) \cdot \ldots \cdot (\binom{n}{m-2} - \ell + 1) \leq n^{\ell m - 2\ell}$ choices for the edges $X_0, \ldots, X_{\ell-1}$. This makes at most $n^{\ell \cdot (m-1)}$ choices altogether.
$\hspace{13cm}$ □

With these observations at hand we can essentially repeat the proof of Theorem 16.1 to also prove Theorem 16.4.

Proof of Theorem 16.4. Let δ be a positive real such that $\ell \cdot \delta < 1$. Let

$$p = n^{1-m+\delta}$$

and consider the random m-uniform hypergraph $H_{n,p}$, where we include each set from $[n]^m$ with probability p, independently. Denote by X the number of cycles of length less or equal than ℓ in $H_{n,p}$. By Observation 16.6 we have that

$$\mathbb{E}[X] \leq \sum_{2 \leq \tilde{\ell} \leq \ell} n^{\tilde{\ell}(m-1)} \cdot p^{\tilde{\ell}} \leq \ell \cdot n^{\ell \cdot \delta}.$$

From Markov's inequality, we deduce that

$$\mathrm{Prob}[X < 2\ell n^{\ell \cdot \delta}] \geq \tfrac{1}{2}. \qquad (16.3)$$

From Observation 16.5 we conclude that

$$\mathrm{Prob}[\exists \Delta : n \to \omega \text{ s.t. } |\mathrm{sel}(\Delta)| \leq n] \qquad (16.4)$$

$$\leq n^n \cdot \binom{\binom{n}{m}}{n} \cdot (1-p)^{cn^m - n}$$

$$\leq n^n \cdot n^{nm} \cdot e^{-p \cdot cn^m + p \cdot n} = o(1),$$

where the last equality again follows from our choice of p.

Hence, putting (16.3) and (16.4) together shows that, for every sufficiently large n, there exists $E \subseteq [n]^m$ with less than $2\ell n^{\ell \cdot \delta}$ cycles of length at most ℓ and such that $|\mathrm{sel}(\Delta) \cap E| \geq n$ for every coloring $\Delta : n \to \omega$. As n is sufficiently large we can assume that $2\ell n^{\ell \cdot \delta} < n$. Thus, deleting one edge from each short cycle yields a selective set system. $\qquad \square$

Having Theorem 16.4 at hand we easily obtain the following partition theorem due to Nešetřil and Rödl (1976a).

Notation. If $\mathcal{F}$ is a family of graphs we denote by $\mathrm{Forb}(\mathcal{F})$ the family of all graphs which do not contain any member from $\mathcal{F}$ as an (induced) subgraph.

Recall that a graph G is two-connected if G cannot be made disconnected by removing one vertex, alternatively, G is two-connected if any two vertices are joined by at least two (internally) vertex-disjoint paths.

Theorem 16.7. *Let $\mathcal{F}$ be a finite family of two-connected graphs, let $F \in \mathrm{Forb}(\mathcal{F})$ and let r be a positive integer. Then there exists a graph $G \in \mathrm{Forb}(\mathcal{F})$ such that $G \xrightarrow{ind} (F)_r^v$.*

Proof. Let ℓ be the largest cardinality of a graph in $\mathcal{F}$. Without loss of generality we may assume that $\ell > r$. Furthermore, let $H = (n, E)$ with $E \subseteq [n]^{v_F}$ be a v_F-uniform hypergraph with chromatic number larger than r and girth larger than ℓ which exists according to Theorem 16.4. Inscribe into each edge in E a copy of F. Since every two edges in E have at most one common vertex, as otherwise they would form a 2-cycle and thus violate the large girth property, every subgraph induced by at most ℓ edges can be disconnected by removing a single vertex. Therefore, any copy of a graph in $\mathcal{F}$ would have to be spread across more than ℓ edges, which cannot be since every graph in $\mathcal{F}$ has at most ℓ vertices. Thus, no copies of graphs in $\mathcal{F}$ occur, and since for every coloring with r colors we have a monochromatic edge in E, the resulting graph has all desired properties. $\qquad \square$

The canonizing version of this partition theorem of vertices has a particular simple form, as it provides a one-to-one or constantly colored result. Essentially the same proof as before then gives the following result of Nešetřil and Rödl (1978b):

Theorem 16.8. *Let $\mathcal{F}$ be a finite family of two-connected graphs and let $F \in \mathrm{Forb}(\mathcal{F})$. Then there exists a graph $G \in \mathrm{Forb}(\mathcal{F})$ which is selective for F, meaning that for every coloring of the vertices of G there exists an F-subgraph which is colored constantly or one-to-one.* $\qquad \square$

16.2 Sparse Ramsey Families

The methods introduced in the previous section can be adapted to prove the existence of sparse Ramsey families, even more, the existence of sparse canonizing Ramsey families.

Notation. For a family $\mathcal{E} \subseteq [n]^m$ and a positive integer k, we denote by $H_k(\mathcal{E})$ the hypergraph which has the k-element subsets of n as vertices and whose edges are given by the sets in $\mathcal{E}$, i.e.,

$$E(H_k(\mathcal{E})) = \{[X]^k \mid X \in \mathcal{E}\}.$$

Theorem 16.9. *Let k, m and ℓ be positive integers. Then, for n sufficiently large, there exists a family $\mathcal{E} \subseteq [n]^m$ such that $H_k(\mathcal{E})$ has girth larger than ℓ and such that for every coloring $\Delta : [n]^k \to \omega$ there exists $Y \in \mathcal{E}$ which is colored canonically, more precisely, there exists a set $J \subseteq k$ such that*

$$\Delta(B) = \Delta(C) \quad \text{if and only if} \quad B : J = C : J$$

holds for all $B, C \in [Y]^k$.

Remark 16.10. With respect to $k = 2$ and colorings $\Delta : [n]^k \to r$ for a fixed positive integer r this result is due to Spencer (1975b). The general case was established by Rödl (1990).

Again, we first derive the analogues to Observations 16.2 and 16.3. For a coloring $\Delta : [n]^k \to \omega$ we denote by $\mathrm{can}(\Delta)$ the set of those $Y \in [n]^m$ which are colored canonically:

$$\mathrm{can}(\Delta) = \{Y \in [n]^m \mid \Delta \lceil [Y]^k \text{ is canonical}\}.$$

Observation 16.11. *There exists $c_0 = c_0(m) > 0$ such that for every coloring $\Delta : [n]^k \to \omega$, where $n \geq n(k, m)$ is sufficiently large, we have $|\mathrm{can}(\Delta)| \geq c_0\, n^m$.*

Proof. Let M be such that $M \to (m)^k_\omega$. Such an M exists by the Erdős-Rado canonizing theorem (Corollary 1.6). Let n be larger than M. For every coloring $\Delta : [n]^k \to \omega$ each $X \in [n]^M$ contributes at least one $Y \in [X]^m$ to $\mathrm{can}(\Delta)$. On the other hand, every $Y \in [X]^m$ is contained in $\binom{n-m}{M-m}$ subsets $X \in [n]^M$. Hence, for every coloring $\Delta : [n]^k \to \omega$ it follows that

$$|\mathrm{can}(\Delta)| \geq \binom{n}{M}\big/\binom{n-m}{M-m} \geq c_0 \cdot n^m,$$

for an appropriately chosen $c_0 > 0$. □

Observation 16.12. *There exists $c_1 = c_1(k, m) > 0$ such that the hypergraph $H_k([n]^m)$ defined by all m-element subsets of n contains at most $c_1\, n^{\ell \cdot (m-k)+k-1}$ cycles of length ℓ.*

Proof. We have $\binom{n}{m} \leq n^m$ possibilities for the first edge. Having the first edge there exist $\binom{m}{k} \leq m^k$ possibilities for choosing the first vertex and, then, $\binom{n-k}{m-k} \leq n^{m-k}$ possibilities for the second edge, and so forth. Finally, the last edge has to intersect the one before it as well as the first one. As these intersections have to be distinct we have at most $\binom{n-k-1}{m-k-1} \leq n^{m-k-1}$ choices. Altogether this gives an upper bound of

$$n^m \cdot (m^k \cdot n^{m-k})^{\ell-2} \cdot m^{2k} \cdot n^{m-k-1} = c_1\, n^{\ell \cdot (m-k)+k-1}$$

for an appropriately chosen $c_1 = c_1(k, m) > 0$. □

Now Theorem 16.9 can be proven following the patterns of the proof of Theorem 16.1, resp., Theorem 16.4.

Proof of Theorem 16.9. Let δ be a positive real such that $\ell \cdot \delta < 1$. Let

$$p = n^{k-m+\delta}$$

and let $\mathcal{E}_{n,p}$ be a random subset of $[n]^m$, where we include each set in $[n]^m$ with probability p, independently. Denote by X the number of cycles of length less or equal than ℓ in $H_k(\mathcal{E}_{n,p})$. Similarly as before, we deduce from Observation 16.12 and Markov's inequality that

$$\text{Prob}[X < 2c_1 \ell n^{k-1+\ell \cdot \delta}] \geq \tfrac{1}{2}. \tag{16.5}$$

From Observation 16.11 we conclude that

$$\text{Prob}[\exists \Delta : [n]^k \to \omega \text{ s.t. } |\text{can}(\Delta)| \leq n^k] \tag{16.6}$$

$$< \binom{n}{k}^{\binom{n}{k}} \cdot \binom{\binom{n}{m}}{n^k} \cdot (1-p)^{c_0 n^m - n^k}$$

$$\leq n^{k n^k} \cdot n^{m n^k} \cdot e^{-p \cdot c_0 n^m + p \cdot n^k} = o(1),$$

where the last equality again follows from our choice of p.

Using that $2c_1 \ell \cdot n^{k-1+\ell \cdot \delta} < n^k$ for all sufficiently large n, the desired result follows from (16.5) and (16.6) with the same arguments as before. □

16.3 Arithmetic Structures

In this section we apply the methods from the previous section to sets of integers carrying an additional arithmetic structure like, typically, arithmetic progressions. Clearly, we can build an m-uniform hypergraph with vertex set n by considering all arithmetic progressions of length m (that are fully contained in n) as edges. The following theorem states that by considering a suitable subset of all m-term arithmetic progressions we can ensure that the girth of the hypergraph corresponding to these edges is large, while we still keep the Ramsey property.

Theorem 16.13. *Let m and ℓ be positive integers. Then there exists a positive integer $n = n(m, \ell)$ and there exists a family $\mathcal{M}$ of arithmetic progressions of length m in $n = \{0, \ldots, n-1\}$ such that girth($\mathcal{M}$) is larger than ℓ but for every coloring $\Delta : n \to \omega$ there exists an arithmetic progression $A \in \mathcal{M}$ of length m such that $\Delta \rceil A$ is constant or one-to-one.*

This is a sparse version of the canonical van der Waerden's theorem (Theorem 6.4). With respect to colorings $\Delta : n \to r$, where r is a fixed positive integer, we get a sparse version of van der Waerden's theorem on arithmetic progressions which is due to Spencer (1975b).

For the proof we first fix some notation. Let $\mathcal{A}_m(n) \subseteq [n]^m$ denote the set of arithmetic progressions of length m within the first n integers, that is,

$$\mathcal{A}_m(n) = \{A \in [n]^m \mid A = \{a, a + b, \ldots, a + (m - 1)b\} \text{ for some } a, b \in [n]\}.$$

Observe, that $|\mathcal{A}_m(n)| = \Theta(n^2)$, as every arithmetic progression is determined by its first two elements. Given a coloring $\Delta : n \to \omega$ we let

$$\text{can}(\Delta) = \{A \in \mathcal{A}_m(n) \mid \Delta \upharpoonright A \text{ is constant or } \Delta \upharpoonright A \text{ is one-to-one}\}.$$

Observation 16.14. *There exists* $c_0 = c_0(m) > 0$ *such that for every coloring* $\Delta : n \to \omega$, *where* $n \geq n(m)$ *is sufficiently large, we have* $|\text{can}(\Delta)| \geq c_0 n^2$.

Proof. From the canonical van der Waerden theorem (Theorem 6.4) we know that there exists M such that every coloring $\Delta : M \to \omega$ contains an m-term arithmetic progression that is either constantly colored or one-to-one. Observe that this implies that for n (much) larger than M and for every coloring $\Delta : n \to \omega$ every M-term arithmetic progression contains an m-term arithmetic progression that is either constantly colored or one-to-one. As $|\mathcal{A}_M(n)| = \Theta(n^2)$ and every m-term arithmetic progression is contained in at most $c = c(m, M)$ many M-term arithmetic progressions, it follows that $|\text{can}(\Delta)| = \Theta(n^2)$, as claimed. $\square$

Observation 16.15. *There exists* $c_1 = c_1(m) > 0$ *such that* $\mathcal{A}_m(n)$ *contains at most* $c_1 n^\ell$ *cycles of length* ℓ.

Proof. Recall that every arithmetic progression A is determined by any two of its elements, provided we know their positions in A. We can thus bound the number of cycles of length ℓ as follows. First choose an arithmetic progression (less than n^2 choices), then choose one of its members (m choices) and a new element and their positions in the new arithmetic progression (less than $n \cdot m^2$ choices). We repeat this for each of the m elements of the cycle, observing that for the closing edge we get only a factor of m instead of n, as we have to choose the element within the first arithmetic progression. Altogether this gives an upper bound of $n^2 \cdot (m^2 \cdot n)^{\ell-2} \cdot m^3 = cn^\ell$ for an appropriately chosen $c = c(m) > 0$. $\square$

Now Theorem 16.13 can be proven following our by now well established pattern.

Proof of Theorem 16.9. Let δ be a positive real such that $\ell \cdot \delta < 1$. Let

$$p = n^{-1+\delta}$$

and let $A_{n,p}$ be a random subset of $\mathcal{A}_m(n)$, where we include each element in $\mathcal{A}_m(n)$ with probability p, independently. Denote by X the number of cycles of length less or equal than ℓ in $A_{n,p}$. Similarly as before, we deduce from Observation 16.15 and Markov's inequality that

$$\text{Prob}[X < 2c_1\ell n^{\ell\cdot\delta}] \geq \tfrac{1}{2}. \tag{16.7}$$

From Observation 16.14 we conclude that

$$\text{Prob}[\exists\Delta : n \to \omega \text{ s.t. } |\text{can}(\Delta)| \leq n] \tag{16.8}$$

$$\leq n^n \cdot \binom{\Theta(n^2)}{n} \cdot (1-p)^{c_0 n^2 - n}$$

$$\leq n^n \cdot n^{\Theta(n)} \cdot e^{-p\cdot c_0 n^2 + p\cdot n} = o(1),$$

where the last equality again follows from our choice of p. The result now follows from (16.7) and (16.8) with the same arguments as before. $\qquad\square$

Similarly, one can prove a sparse version of the finite sum theorem (Theorem 2.12), resp., its canonizing counterpart (Theorem 6.8).

Notation. For positive integers $m < n$ we denote by SUM_m^n the set of all families $\{\sum_{i\in I} x_i \mid \emptyset \neq I \subseteq m\}$ where $x_0, \ldots, x_{m-1}$ are elements of $[1, n-1]$ such that also $x_0 + \ldots + x_{m-1} < n$.

Theorem 16.16. *Let m and ℓ be positive integers. Then there exists a positive integer $n = n(m, \ell)$ and a family $\mathcal{E} \subseteq \text{SUM}_m^n$ with $\text{girth}(\mathcal{E}) > \ell$ such that for every coloring $\Delta : n \to \omega$ there exists $S = \{\sum_{i\in I} x_i \mid \emptyset \neq I \subseteq m\} \in \mathcal{E}$ such that $\Delta\rceil S$ is canonical.*

Also a sparse version of Deuber's partition theorem for (m, p, c)-sets (Theorem 2.11) can be established using similar arguments.

Theorem 16.17. *Let m, p, c, ℓ and r be positive integers. Then there exists a family $\mathcal{M}$ of (m, p, c)-sets with $\text{girth}(\mathcal{M}) > \ell$ such that for every r-coloring $\Delta : \mathbb{N} \to r$ of the positive integers there exists a monochromatic (m, p, c)-set in $\mathcal{M}$.*

Corollary 16.18. *Let ℓ and r be positive integers and let $A \cdot x = 0$ be a partition regular system of linear equations. Then there exists a family S of solutions of $A \cdot x = 0$ (where a solution is viewed as a subset of $\mathbb{N}$) with $\text{girth}(S) > \ell$ and such that for every r-coloring $\Delta : \mathbb{N} \to r$ there exists a monochromatic solution.*

This strengthens, e.g., Theorem 16.13 in the sense that also the difference of the corresponding arithmetic progression is included. These results are from Ruciński, Voigt (unpublished).

16.4 Parameter Sets

In this section we consider a sparse version of the Hales-Jewett theorem
(Theorem 4.2). To state the result precisely let us first fix some notation. Given
an alphabet A of size $t = |A|$ and integers m, n we define a hypergraph $\mathcal{H}(n)$ for
m-parameter words similarly as in Chap. 14. More precisely, the vertex set consists
of all words of length n over A, i.e., $V(\mathcal{H}_k(n)) = A^n = [A]\binom{n}{0}$, while the edges
correspond to the m-parameter words in $[A]\binom{n}{m}$. That is, for every $f \in [A]\binom{n}{m}$ the
hyperedge e_f is given by

$$e_f = \{f \cdot g \mid g \in [A]\binom{m}{0}\}.$$

Note that the hypergraph $\mathcal{H}(n)$ also depends on the alphabet A and the parameter
size m; this is not shown in the notation as A and m will be viewed as fixed
throughout this section. With this notation at hand we can now state the sparse
Hales-Jewett theorem.

Theorem 16.19. *Let A be an alphabet of size $t = |A|$ and let m, ℓ and r be positive
integers. Then there exists a positive integer $n = n(t, m, \ell, r)$ and a subhypergraph
$\mathcal{H}' \subseteq \mathcal{H}(n)$ with $\mathrm{girth}(\mathcal{H}') > \ell$ such that for every coloring $\Delta : A^n \to r$ there exists
an $e_f \in E(\mathcal{H}')$ such that $\Delta_f : A^m \to r$ given by $\Delta_f(g) = \Delta(f \cdot g)$ is a constant
coloring.*

This result was first proved by Rödl (1990) using different methods. We will
prove Theorem 16.19 following the technique that we established in the previous
sections. To make our life easier we apply the probabilistic method only to a
subgraph of $\mathcal{H}(n)$. This subgraph is defined as follows. Let $M = HJ(t, m, r)$
be according to the Hales-Jewett theorem (Theorem 4.2). Then $\mathcal{H}_M(n) \subseteq \mathcal{H}(n)$
consists of those edges which contain at most M parameters. More precisely, for
$f = (f_0, \ldots, f_{n-1}) \in [A]\binom{n}{m}$ let Π_f denote the positions in f that consist of a
parameter, that is

$$\Pi_f = \{i \in n \mid f_i \notin A\} \quad \text{and} \quad \pi_f = |\Pi_f|.$$

The hypergraph $\mathcal{H}_M(n)$ consists then of those edges e_f for which $\pi_f \leq M$. That is,

$$E(\mathcal{H}_M(n)) = \{e_f \mid f \in [A]\binom{n}{m} \text{ and } \pi_f \leq M\}.$$

We call an edge e_f of type i if $\pi_f = i$. Observe that the number of edges of type i
is bounded from below by $\binom{n}{i} \cdot \binom{i}{m} \cdot t^{n-i}$ and from above by $\binom{n}{i} \cdot m^i \cdot t^{n-i}$. As we
only consider edges of type at most M, where $M = HJ(t, m, r)$ does not depend
on n, this implies that there exists a constant $c_e = c_e(t, m, r)$ such that the number
of edges of type i is at most $c_e n^i \cdot t^n$.

Observation 16.20. *There exists $c_0 = c_0(t,m,r) > 0$ such that for every coloring $\Delta : A^n \to r$, where n is sufficiently large, there exists $m \leq i \leq M$ so that we have at least $c_0 n^i t^n$ monochromatic edges of type i.*

Proof. Consider an M-parameter word $g \in [A]\binom{n}{M}$ in which every parameter occurs only once, i.e., $\pi_g = M$. Observe that Δ and g induce a coloring $\Delta_g : A^M \to r$. By choice of M (and the classical Hales-Jewett theorem) we deduce that for every such g there exists an $f \in [A]\binom{M}{m}$ that is monochromatic. Note that this implies that $f_g := g \cdot f \in [A]\binom{n}{m}$ is monochromatic with respect to Δ and of type $\pi_{f_g} \leq M$. Note also that we get the same f_g for at most $\binom{n - \pi_{f_g}}{M - \pi_{f_g}}$ many g's.

For a contradiction, assume that the claim is false. Then we have

$$t^{n-M} \cdot \binom{n}{M} = |\{g \mid g \in [A]\binom{n}{M} \text{ s.t. } \pi_g = M\}| \leq \sum_{i=m}^{M} c_0 n^i t^n \cdot \binom{n-i}{M-i},$$

which is easily seen to be false for n sufficiently large and an appropriately chosen $c_0 = c_0(t,m,r)$. (Recall that $M = M(t,m,r)$ is a function of t, m, and r, but independent of n.) $\qquad\square$

Recall that a cycle of length ℓ in $\mathcal{H}_M(n)$ consists of ℓ pairwise different edges $e_0, \ldots, e_{\ell-1}$ and ℓ pairwise different vertices $x_0, \ldots, x_{\ell-1}$ such that $x_i \in e_i \cap e_{i+1}$ for $i \leq \ell - 2$ and $x_{\ell-1} \in e_{\ell-1} \cap e_0$. We say a cycle is of type i_0, if $\sum_{i \in \ell} \pi_{e_i} = i_0$. Note that by definition of $\mathcal{H}_M(n)$ a cycle of length ℓ can have type at most $\ell \cdot M$.

Observation 16.21. *There exists a $c_1 = c_1(t,m,\ell,r) > 0$ such that for all $m\ell \leq i_0 \leq \ell M$ we have: $\mathcal{H}_M(n)$ contains at most $c_1 n^{i_0-1} t^n$ cycles of length ℓ and type i_0.*

Proof. For a cycle $e_0, \ldots, e_{\ell-1}$ consider the number of coordinates so that at least one of the edges e_i contains a parameter in this coordinate. Clearly, this number is at most i_0. Assume, for a contradiction, that it is exactly i_0. Then the sets Π_{e_i} are pairwise disjoint. I.e., for every coordinate at most one of the edges e_i contains a parameter. Note that this implies (due the definition of a cycle) that all the other edges must *all* have the *same* letter at this coordinate. Observe that this in turn implies that every pair of edges e_i, e_{i+1} of edges intersects in the *same* vertex, which contradicts the fact that the vertices x_i need to be pairwise different.

Hence, we see that we need to have $\sum \pi_{e_i} \leq i_0 - 1$. We can thus choose the set of coordinates that may contain a parameter ($\binom{n}{i_0-1}$ choices) choose the letters for the remaining positions. (Note: these have to be identical for all edges; thus there are just t^{n-i_0+1} choices to do that) And finally decide for each edge which value it takes on the special coordinates: at most $(t+m)^{(i_0-1)\cdot\ell}$ choices. As $i_0 \leq \ell M$ and M is bounded by a function in t, m, and r this concludes the proof. $\qquad\square$

Now Theorem 16.19 can be proven following our by now well established pattern.

Proof of Theorem 16.19. We choose a random subset of H_n of $\mathcal{H}_M(n)$ as follows. For $i \in [m, M]$ we define

$$p_i = n^{-i+\delta},$$

where δ is a positive real such that $\ell \cdot \delta < 1$, and include all edges of type i in $\mathcal{H}_M(n)$ with probability p_i, independently.

Denote by X the number of cycles of length less or equal than ℓ in H_n. In order to calculate the expectation of X we distinguish the types of the edges within the cycle (so that we can calculate the probability of the appearance of a cycle). Note that this also determines the type of a cycle and we can thus apply Observation 16.21.

$$\mathbb{E}[X] \leq \sum_{2 \leq \tilde{\ell} \leq \ell} \sum_{i_0,\dots,i_{\tilde{\ell}-1} \in [m,M]} c_1\, n^{\sum_{j \in \tilde{\ell}} i_j - 1}\, t^n \cdot \prod_{j \in \tilde{\ell}} p_{i_j} \leq c_1' \cdot t^n \cdot n^{-1+\ell\delta}.$$

for an appropriately chosen constant $c_1' = c_1'(t, m, \ell, r)$ (recall that M is a constant depending on t, m, and r, but not on n.) Markov's inequality thus implies for all n sufficiently large:

$$\mathrm{Prob}[X < \tfrac{1}{2}t^n] \geq \mathrm{Prob}[X < 2c_1' t^n n^{-1+\ell\delta}] \geq \tfrac{1}{2}. \tag{16.9}$$

From Observation 16.20 we conclude that for every r-coloring $\Delta : A^n \to \omega$ there exits an $i \in [m, M]$ such that there exit at least $c_0 n^i t^n$ monochromatic edges of type i. The probability that less than t^n of these are present in H_n can be bounded by

$$\binom{c_e\, n^i t^n}{t^n} \cdot (1 - p_i)^{c_0 n^i t^n - t^n} \leq (e c_e\, n^i)^{t^n} e^{-p_i(c_0 n^i t^n - t^n)} \leq e^{-\frac{1}{2}c_0 n^\delta t^n},$$

for all n sufficiently large. From the union bound we thus get

$$\mathrm{Prob}[\exists \Delta : [A]\binom{n}{k} \to \omega \text{ s.t. } \leq t^n \text{ edges in } H_n \text{ are monochromatic}]$$

$$\leq r^{t^n} \cdot M \cdot e^{-\frac{1}{2}c_0 n^\delta t^n} = o(1). \tag{16.10}$$

The result now follows from (16.9) and (16.10) with the same arguments as before. $\qquad\square$

Apparently it is not that easy to establish a sparse Graham-Rothschild theorem using the probabilistic method. Such a result was established by Prömel and Voigt (1988) using more involved deterministic constructions. We do not cover that result here. Instead we consider another natural extension of the sparse Hales-Jewett theorem, namely a sparse version of the canonical Hales-Jewett theorem (Theorem 6.1).

Theorem 16.22. *Let A be an alphabet of size $t = |A|$ and let m and ℓ be positive integers. Then there exists a positive integer $n = n(t, m, \ell)$ and a subhypergraph*

$\mathcal{H}' \subseteq \mathcal{H}(n)$ with girth$(\mathcal{H}') > \ell$ such that for every coloring $\Delta : A^n \to \omega$ there exists an $e_f \in E(\mathcal{H}')$ and an equivalence relation $\approx$ on A such that for all $g, h \in [A]\binom{m}{0}$ it follows that

$$\Delta(f \cdot g) = \Delta(f \cdot h) \quad \text{if and only if} \quad g/\approx \; = \; h/\approx,$$

$$\text{i.e., } g(i) \approx h(i) \text{ for every } i < m.$$

Reviewing the proof of the sparse Hales-Jewett theorem, we see that the general setup of the proof would also work for the canonical case: we could just replace the definition of $M = HJ(t, m, r)$ by $M = CHJ(t, m, r)$, where $CHJ(..)$ is the function defined in the canonical Hales-Jewett theorem (Theorem 6.1), and then everything would work as before. Unfortunately, there is one catch where this approach fails. In the canonical case we need to consider unbounded colorings $\Delta : A^n \to \omega$ instead of bounded colorings $\Delta : A^n \to r$. And the number of unbounded colorings is t^{nt^n}, which is too large for the union bound argument in the last part of the proof of Theorem 16.19.

In the remainder of this section we provide an easy fix for this problem. The main idea is to consider words of length $\hat{n}$ that consist of $N = N(\ell)$ blocks of length n, where n is large enough so that the argument of the previous section works for this value, meaning that the inequality in Eq. (16.10) holds for n (where M is chosen as in the canonical theorem). That is, we define a hypergraph $\mathcal{H}_M(N, n)$ as follows. Vertices are words of length $N \cdot n$ over the alphabet A. Edges correspond to m-parameter words $f \in [A]\binom{Nn}{m}$ – but we only consider those f's that have the property that in each block we have at most M occurrences of a parameter and, in addition, each parameter occurs the same number of times in each block. Note that this implies in particular that each parameter appears at least once per block and that the number of parameters is identical in all blocks. Similarly as before, we call this number of occurrences of parameters per block the *type* of an edge. Observe there exists a constant $c_e = c_e(t, m, r)$ such that the number of edges of type i is bounded by $[c_e \, n^i \cdot t^n]^N$.

With these definitions at hand we can now rephrase the observations from above.

Observation 16.23. *There exists $c_0 = c_0(t, m, N) > 0$ such that for every coloring $\Delta : A^{Nn} \to \omega$, where n is sufficiently large, there exists $m \le i \le M$ so that we have at least $c_0[n^i t^n]^N$ edges of type i which induce a canonical coloring.*

Proof. We proceed similarly as in the proof of Observation 16.20. Consider an M-parameter word $g \in [A]\binom{Nn}{M}$ in which every parameter occurs exactly only once in each block. Observe that Δ and g induce a coloring $\Delta_g : [A]\binom{M}{0} \to \omega$. By choice of M (and the canonical Hales-Jewett theorem) we deduce that for every such g there exists an $f \in [A]\binom{M}{m}$ that induces a canonical coloring. Note that this implies that $f_g := g \cdot f \in [A]\binom{Nn}{m}$ is canonical with respect to Δ and of type $\pi_{f_g} \le M$. Note also that f_g has the desired properties, namely that each parameter occurs the same number of times in each block. Finally, observe that we get the same f_g for at most $[\binom{n - \pi_{f_g}}{M - \pi_{f_g}}]^N$ many g's.

For a contradiction, assume that the claim is false. Then we have

$$[t^{n-M} \cdot \binom{n}{M}]^N = |\{g \mid g \in [A]\binom{n}{M} \text{ s.t. } \pi_g = M \text{ per block}\}|$$

$$\leq \sum_{i=m}^{M} c_0 [n^i t^n \cdot \binom{n-i}{M-i}]^N ,$$

which is easily seen to be false for n sufficiently large and an appropriately chosen $c_0 = c_0(t, m, N)$. (Recall that $M = M(t, m)$ is a function of t and m, but independent of n.).   □

We say that the type of a cycle is equal to the sums of the types of its edges. Note that by definition of $\mathcal{H}_M(N, n)$ a cycle of length ℓ can have type at most $\ell \cdot M$.

Observation 16.24. *There exists a $c_1 = c_1(t, m, \ell, N) > 0$ such that for all $m\ell \leq i_0 \leq \ell M$ we have: $\mathcal{H}_M(N, n)$ contains at most $c_1 [n^{(i_0-1)} t^n]^N$ cycles of length ℓ and type i_0.*

Proof. Recall the following fact from Observation 16.20. If for some coordinate exactly one of the edges contains a parameter, say λ_i, then this implies that all other edges must have the *same* letter at this coordinate. Note also that this implies that in all vertices that lie in the intersection of two edges this coordinate has to be fixed to this letter, which implies that the parameter λ_i can take only this value. As we need to have ℓ pairwise different vertices in the intersection of the edges, this implies that at least one parameter, say $\lambda_{\hat{i}}$, has the property that all its appearances are such that there always exists at least one more parameter in the same coordinate in one of the other edges of the cycle.

Assume we have fixed all the occurrences of the parameters in each of the edges $e_0, \ldots, e_{\ell-2}$. Then all the occurrences of the parameter $\lambda_{\hat{i}}$ in edge $e_{\ell-1}$ can only happen at positions of parameters within the edges $e_0, \ldots, e_{\ell-2}$. As parameter $\lambda_{\hat{i}}$ occurs at least once per block, we thus deduce that the total number of coordinates where at least one of the edges contains a parameter is bounded by $N i_0 - N$. The total number of cycles can thus be bounded as follows. First we choose the set of coordinates that may contain a parameter ($\binom{n}{i_0-N}$ choices), then we choose the letters for the remaining positions ($t^{N(n-i_0+1)}$ choices). And finally we decide for each edge which value it takes in the special coordinates: at most $(t + m)^{(Ni_0-N)\cdot\ell}$ choices. As $i_0 \leq \ell M$ and M is bounded by a function in t, m, and r this concludes the proof.   □

Now a canonical sparse Hales-Jewett theorem can be proven as before.

Proof of Theorem 16.22. We choose a random subset H_n of $\mathcal{H}_M(n, N)$ as follows. For $i \in [m, M]$ we define

$$p_i = n^{-(N \cdot (i+\delta))},$$

where δ is a positive real such that $\ell \cdot \delta < 1$, and include all edges of type i in $\mathcal{H}_M(n, N)$ with probability p_i, independently.

Denote by X the number of cycles of length less or equal than ℓ in H_n. In order to calculate the expectation of X we distinguish the types of the edges within the cycle (so that we can calculate the probability of the appearance of a cycle). Note that this also determines the type of a cycle and we can thus apply Observation 16.24.

$$\mathbb{E}[X] \leq \sum_{2 \leq \tilde{\ell} \leq \ell} \sum_{i_0, \dots, i_{\tilde{\ell}-1} \in [m, M]} c_1 \, n^{N \cdot \sum_{j \in \tilde{\ell}} i_j - N} t^{Nn} \cdot \prod_{j \in \tilde{\ell}} p_{i_j} \leq c_1' \cdot t^{Nn} \cdot n^{-N + N\ell\delta}.$$

for an appropriately chosen constant $c_1' = c_1'(t, m, \ell, r)$ (recall that M is a constant that depends only on t, m, and r.) Markov's inequality thus implies for all n sufficiently large:

$$\text{Prob}[X < \tfrac{1}{2} t^{Nn}] \geq \text{Prob}[X < 2c_1' \, t^{Nn} n^{-N + N\ell\delta}] \geq \tfrac{1}{2}. \qquad (16.11)$$

From Observation 16.23 we conclude that for every r-coloring $\Delta : A^{Nn} \to \omega$ there exits an $i \in [m, M]$ such that there exits at least $c_0[n^i t^n]^N$ edges of type i which induce a canonical coloring. The probability that less than t^{Nn} of these are present in H_n can be bounded by

$$\binom{[c_e \, n^i t^n]^N}{t^{nN}} \cdot (1 - p_i)^{c_0 \, [n^i t^n - t^n]^N} \leq (e c_e \, n^{iN}) t^{nN} e^{-p_i (c_0 \, [n^i t^n]^N - t^{nN})}$$

$$\leq e^{-\frac{1}{2} c_0 [n^\delta t^n]^N},$$

for all n sufficiently large. From the union bound we thus get

$$\text{Prob}[\exists \Delta : A^{Nn} \to \omega \text{ s.t. } \leq t^{Nn} \text{ edges in } H_n \text{ are canonical}] \qquad (16.12)$$

$$\leq t^{nN t^{nN}} \cdot e^{-\frac{1}{2} c_0 [n^\delta t^n]^N} = o(1),$$

if we choose N such that $\delta N > 1$. The result now follows from (16.10) and (16.11) with the same arguments as before. $\square$

Part V
Density Ramsey Theorems

Chapter 17
Szemerédi's Theorem

In 2012 Endré Szemerédi, born 1940 in Budapest, received the Abel prize in mathematics. Together with the fields medal the Abel prize is the most prestigious award in mathematics. It is often described as the mathematician's Nobel prize. From the press release in 2012:

> Many of his discoveries carry his name. One of the most important is Szemerédi's Theorem, which shows that in any set of integers with positive density, there are arbitrarily long arithmetic progressions. Szemerédi's proof was a masterpiece of combinatorial reasoning, and was immediately recognized to be of exceptional depth and importance. A key step in the proof, now known as the Szemerédi Regularity Lemma, is a structural classification of large graphs.

Let us recap Szemerédi's theorem in more depth. In 1936 Erdős and Turán conjectured the following famous generalization of van der Waerden's theorem:

Let $\epsilon > 0$ and k be a positive integer. Then there exists a positive integer $n = n(k, \epsilon)$ such that every set $S \subseteq n$ satisfying $|S| \geq \epsilon n$ contains an arithmetic progression of length k.

In 1953 Roth proved this conjecture for $k = 3$ using analytic number theory. Szemerédi (1969) extended Roth's result to 4-element progressions and, finally, in 1975 he was able to settle the conjecture in its full generality. Szemerédi's proof is a combinatorial masterpiece – that earned him a place in history. Later, Furstenberg (1977) gave a different proof using techniques from ergodic theory and, applying similar methods, Furstenberg and Katznelson (1978) extended Szemerédi's theorem to higher dimensions, proving a density version of Gallai-Witt's theorem. By now, Szemerédi's original combinatorial proof and Fürstenberg's ergodic proof (cf. also Furstenberg et al. 1982) are not the only approaches known. In 2001 Gowers provided a new proof using Fourier-analytic methods, and Gowers (2006, 2007), Rödl and Skokan (2004) and Nagle et al. (2006) obtained proofs using hypergraph removal lemmas.

It is the great achievement of progress in science that results that deserve the highest prizes that a community awards will eventually become 'common knowledge'. The proof of the Polymath-project on the density Hales-Jewett theorem

H.J. Prömel, *Ramsey Theory for Discrete Structures*,
DOI 10.1007/978-3-319-01315-2__17,
© Springer International Publishing Switzerland 2013

achieved exactly that for Szemerédi's theorem: nowadays it can be proved 'on a few pages' by 'elementary means'. The aim of this chapter is to provide such a proof.

Actually, however, we will not be concerned with Szemerédi's theorem directly. As we have seen in previous chapters of this book, it sometimes simplifies matters if we study problems in a more general setting. This is what we do here as well. All we need is an appropriate mapping from A^n to $\mathbb{N}$. For example, let $A = \{0, \ldots, k-1\}$ and consider the bijection $\kappa : A^n \to k^n$ given by

$$\kappa(a_0, \ldots, a_{n-1}) = \sum_{i<n} a_i k^i.$$

Then for every combinatorial line $f \in [A]\binom{n}{1}$ we have that

$$\{\kappa(f \cdot i) \mid i < k\}$$

is a k-term arithmetic progression. In order to prove Szemerédi's theorem it thus suffices to find combinatorial lines in sparse subsets of combinatorial spaces A^n. This is what we will do in the next chapter.

Chapter 18
Density Hales-Jewett Theorem

Van der Waerden's theorem guarantees the existence of a monochromatic arithmetic progression. While historically it was proven as a result by itself, nowadays we easily obtain it as a special case of Hales-Jewett's theorem, cf. Chap. 4. In the light of Szemerédi's theorem it is thus very natural to ask for a density version of Hales-Jewett's theorem for lines:

Let $\epsilon > 0$ and A be a finite alphabet. Then there exists a positive integer $n = n(|A|, \epsilon)$ such that every set $S \subseteq A^n$ satisfying $|S| \geq \epsilon |A|^n$ contains a combinatorial line, i.e., there exists $f \in [A]\binom{n}{1}$ so that $f \cdot A \subseteq S$.

This result was proven in 1991 by Furstenberg and Katznelson using ergodic methods. As we will see below, their result actually implies the result for d-spaces as well.

In 2009 Tim Gowers started an experiment on his blog suggesting to the community to collaboratively try to solve an important mathematical problem by posting comments on his blog. As a topic for the first problem to be solved in this way he suggested to find a combinatorial proof of the Fürstenberg and Katznelson result. Already 7 weeks later the project was completed and such a proof found; it was published under a pseudonym, cf. Polymath (2012).

Theorem 18.1 (Density Hales-Jewett Theorem). *Let $\delta > 0$, $d \geq 1$ an integer, A be a finite alphabet. Then there exists a positive integer $n = DHJ(|A|, d, \delta)$ such that every set $S \subseteq A^n$ satisfying $|S| \geq \delta |A|^n$ contains a combinatorial d-space, i.e., there exists $f \in [A]\binom{n}{d}$ so that $f \cdot A^d \subseteq S$.*

A few years after the polymath project was completed Dodos et al. (2013) provided a simplified version of the Polymath-approach. In this chapter we will mostly follow their approach.

H.J. Prömel, *Ramsey Theory for Discrete Structures*,
DOI 10.1007/978-3-319-01315-2__18,
© Springer International Publishing Switzerland 2013

18.1   Lines Imply Spaces

It was known long before Fürstenberg and Katznelson proved the density results
for lines that this result actually implies the density Hales-Jewett theorem in its full
generality. This was shown by Brown and Buhler (1984) in a paper called "Lines
imply spaces in density Ramsey theory":

Lemma 18.2. *Let $k \geq 2$ and assume Theorem 18.1 is true for alphabets of size k
and $d = 1$. Then it also holds for alphabets of size k and all $d \geq 2$.*

Proof. Let A be an alphabet of size $|A| = k$. We prove the result by induction on
d. The case $d = 1$ is handled by the assumption of the lemma. So assume now
Theorem 18.1 is true for alphabets of size k and all $d \leq d_0$ for some $d_0 \geq 1$ (and
all $\delta > 0$). We show that it then also holds for $d_0 + 1$ (again, for all $\delta > 0$).

Fix some $\delta > 0$ arbitrarily and let $N = DHJ(k, d_0, \frac{\delta}{2})$. Put $t = (k + d_0)^N$ and
observe that t is an upper bound on the number of d_0-parameter words of length N
over A. Choose $M = DHJ(k, 1, \frac{\delta}{2t})$. Observe that M and N exist by our induction
assumption.

Now let $S \subseteq A^{M+N}$ so that $|S| \geq \delta |A|^{M+N}$. We claim that there exists $f \in
[A]\binom{n}{1+d_0}$ that is completely contained in S, i.e., that satisfies $f \cdot A^{1+d_0} \subseteq S$.

For $g \in A^M$ let $S_g = \{h \in A^N \mid g \times h \in S\}$. Moreover, let $I \subseteq A^M$ be the set
of those initial segments which have sufficiently many tails in S, i.e.,

$$I = \{g \in A^M \mid |S_g| \geq \tfrac{\delta}{2} k^N\}.$$

Then, by choice of N, for every $g \in I$ there exists $f_g \in [A]\binom{N}{d_0}$ such that $g \times f_g$ is
completely contained in S, i.e., $(g \times f_g) \cdot A^{d_0} \subseteq S$.

Observe that $|I| \geq \frac{\delta}{2} k^M$. Otherwise the total number of elements in S would be
less than

$$\tfrac{\delta}{2} k^M \cdot k^N + (1 - \tfrac{\delta}{2}) k^M \cdot \tfrac{\delta}{2} k^N = \delta(1 - \tfrac{\delta}{4}) k^{M+N} < \delta k^{M+N},$$

a contradiction.

By the pigeonhole principle there are at least $\frac{\delta}{2t} k^M$ many $g \in I$ which give rise
to the same f_g, say to $f_N \in [A]\binom{N}{d_0}$.

But now, by choice of $M = DHJ(k, 1, \frac{\delta}{2t})$, there exists $f_M \in [A]\binom{M}{1}$ such that
$f_M \times f_N \in [A]\binom{M+N}{1+d_0}$ is completely contained in S. $\square$

18.2   The Boolean Case: Sperner's Lemma

Recall from Sect. 3.1.3 that every element $f \in [2]\binom{n}{0}$ can be interpreted as the
characteristic function of a subset of $n = \{0, \ldots, n-1\}$, i.e. as an element of the
Boolean lattice $\mathcal{B}(n)$. A combinatorial line $\ell \in [2]\binom{n}{1}$ then corresponds to a chain of
length 1 in $\mathcal{B}(n)$, i.e. to two sets $A, B \subseteq n$ such that $A \subsetneq B$.

The only substructures of $\mathcal{B}$ that do not contain a combinatorial line are thus antichains. The size of a largest antichain was determined by Sperner (1928).

Theorem 18.3 (Sperner's Lemma).   *Every antichain in the boolean lattice* $\mathcal{B}(n)$ *has size at most* $\binom{n}{\lfloor n/2 \rfloor}$.

Note that an antichain of size $\binom{n}{\lfloor n/2 \rfloor}$ is obtained by taking all sets of size exactly equal to $\lfloor n/2 \rfloor$.

Proof. We choose a random element from $\mathcal{B}(n)$ as follows: first we choose a permutation π of the n elements uniformly at random, then we choose an integer $m \in \{0, \ldots, n\}$ uniformly at random and return the set R consisting of the first m elements according to the permutation, i.e., $R = \{\pi(0), \ldots, \pi(m-1)\}$.

As $\mathcal{A}$ is an antichain we know that for every permutation at most one m will result in a set from $\mathcal{A}$. Thus, $\text{Prob}[R \in \mathcal{A}] \leq \frac{1}{n+1}$.

Now we look at this random experiment from a different angle: consider a set A of size, say, $|A| = i$. What is the probability that $R = A$? – Clearly, m has to be chosen as i (with probability $1/(n+1)$) and the permutation π has to be such that the first i elements are the elements from A. This happens with probability $i! \cdot (n-i)!/n! = \binom{n}{i}^{-1}$.

Fix reals δ_i that correspond to the fraction of subsets of size i that belong to $\mathcal{A}$, i.e.. $|\mathcal{A}| = \sum_{i \leq n} \delta_i \binom{n}{i}$. Then our two observations from above imply

$$\frac{1}{n+1} \geq \text{Prob}[R \in \mathcal{A}] = \sum_{i \leq n} \delta_i \binom{n}{i} \cdot \frac{1}{n+1} \binom{n}{i}^{-1},$$

from which we deduce that $\sum_{i \leq n} \delta_i \leq 1$. Thus, $\mathcal{A}$ has maximum size if it contains all i-element sets, where i is chosen such that $\binom{n}{i}$ is maximized. I.e, $i = \lfloor n/2 \rfloor$ or $i = \lceil n/2 \rceil$, as claimed. $\square$

Using Stirling's formula one obtains that

$$\binom{n}{\lfloor n/2 \rfloor} \approx \sqrt{\frac{2}{\pi}} \frac{2^n}{\sqrt{n}} = o(2^n),$$

thus proving Theorem 18.1 for the case $k = 2$ and $d = 1$ (and all $\delta > 0$). Using the 'lines imply spaces' result of the previous section we immediately deduce:

Corollary 18.4.   *Theorem 18.1 holds for alphabets of size* $k = 2$ *for all* $d \geq 1$ *and all* $\delta > 0$. $\square$

18.3   The General Case: Alphabets of Size $k \geq 3$

We will prove Theorem 18.1 by induction on k, the size of the alphabet. Sperner's Lemma, cf. previous section, settles the base case. From now on we thus assume that Theorem 18.1 holds for some $k \geq 2$ (for all $d \geq 1$ and all $\delta > 0$). We will prove that it then also holds for all alphabets of size $k + 1$. Recall that the fact that 'lines imply spaces' (Sect. 18.1) implies that we only have to consider the case $d = 1$, that is, we only have to show that for every $\delta > 0$ there exists n sufficiently large (as a function of $k + 1$ and δ) such that all subsets $S \subseteq A^n$ of density δ contain a combinatorial line.

The key tool for achieving this is the following lemma that states that for a set S of given density we will either find a combinatorial line in S or there exist a subspace (of suitable dimension) in which S has a higher density. Formally,

Lemma 18.5 (Key Lemma). *Let $k \geq 2$ and assume Theorem 18.1 is true for alphabets of size k. Then there exists for every $\delta > 0$ an $m_0 = m_0(k + 1, \delta)$ and $\sigma = \sigma(k + 1, \delta)$ such that for every $m \geq m_0$ there exists $n_0 = KL(k + 1, m, \delta)$ such that the following is true. If $|A| = k + 1$, $n \geq n_0$, and $S \subseteq A^n$ satisfies $\delta_s := |S|/|A|^n > \delta$ then at least one of the following two properties hold:*

(1) There exists $\ell \in [A]\binom{n}{1}$ such that $\ell \cdot A \subseteq S$, or
(2) There exists $f \in [A]\binom{n}{m}$ such that $|S \cap f \cdot A^m| \geq (\delta_s + \sigma)|A|^m$.

With this lemma at hand the proof of Theorem 18.1 is easily completed.

Proof of Theorem 18.1. Let A be an alphabet of size $k+1$ and assume Theorem 18.1 holds for alphabets of size k. Recall that it suffices to consider the case $d = 1$, cf. Lemma 18.2. Fix $\delta > 0$ arbitrarily. Here is the idea. Given $S \subseteq A^n$ we apply Lemma 18.5. Either we find a line contained in S, in which case we are done, or we get a subspace $f \in [A]\binom{n}{m}$ in which S has higher density. If m is large enough we can apply Lemma 18.5 again, this time with respect to A^m and $S' := \{x \in A^m \mid f \cdot x \in S\}$. Again, we either find a line, in which case we are done, or we get a subspace $f \in [A]\binom{n}{m}$ in which S has higher density. As the density increases by σ in each step (observe that σ does not depend on m!) this process can only continue for at most $\lceil \sigma^{-1} \rceil$ rounds. All we thus have to ensure is that we start with a large enough n (and suitably defined m's) so that the process can run that long. This is what we now do.

Let $m_{\lceil \sigma^{-1} \rceil + 1} := m_0$ and define $m_i := KL(k + 1, m_{i+1}, \delta)$ for all $i = \lceil \sigma^{-1} \rceil, \ldots, 0$, where $KL(\cdot)$ is the function provided by Lemma 18.5. Let $DHJ(k + 1, 1, \delta) := m_0$. By applying Lemma 18.5 to a set $S_0 \subset A^n$ with density $\delta_s := |S_0|/|A^n| \geq \delta$, for some $n \geq m_0$, we either find the desired line in S_0 or a subspace $f_1 \in [A]\binom{n}{m_1}$ in which S_0 has density at least $\delta_s + \sigma$. Let $S_1 := \{x \in A^{m_1} \mid f_1 \cdot x \in S\}$ and continue as outlined above. In the jth application of Lemma 18.5 we either find a combinatorial line $\ell \in [A]\binom{m_{j-1}}{1}$ contained in S_{j-1} (in which case $f_1 \cdot \ldots \cdot f_{j-1} \cdot \ell$ is contained in S_0, as desired) or a subspace $f_j \in [A]\binom{m_{j-1}}{m_j}$ in which S_{j-1} has density at least $\delta_s + j \cdot \sigma$. Setting $S_j := \{x \in A^{m_j} \mid f_j \cdot x \in S_{j-1}\}$ thus allows for a $(j + 1)$st application of Lemma 18.5. As no set can have a density greater than one, this process has to yield a combinatorial line after at most $\lceil \sigma^{-1} \rceil$ rounds, and the proof of Theorem 18.1 is completed.   □

In the remainder of this section we prove Lemma 18.5. While doing so we always think of A as an alphabet of size $k + 1$. We assume that A consists of the letters $a_0, \ldots, a_k$. By removing the letter a_k we obtain an alphabet of size k (to which we can thus apply the induction assumption that Theorem 18.1 holds for alphabets of size k for all $d \geq 1$ and all $\delta > 0$). We use B to denote this restricted alphabet, i.e, $B = A \setminus \{a_k\} = \{a_0, \ldots, a_{k-1}\}$.

For the proof of Lemma 18.5 we proceed in three steps. First we show that we either find a subspace in which S is denser or we find a subspace in which S has almost the same density, but in addition at least some fraction of all lines that do not use the letter a_k lie completely in S. Formally,

Lemma 18.6. *Let $k \geq 2$ and assume Theorem 18.1 is true for alphabets of size k. Then there exists for every $\delta > 0$ an $m_0 = m_0(k + 1, \delta)$ and $\rho = \rho(k + 1, \delta) \leq \frac{1}{2}\delta$ such that for every $m \geq m_0$ and there exists an $N_1 = N_1(k + 1, m, \delta)$ such that the following is true. If $|A| = k + 1$, $n \geq N_1$, and $S \subseteq A^n$ satisfies $\delta_s := |S|/|A|^n > \delta$ then at least one of the following two properties hold:*

(1) There exists $f \in [A]\binom{n}{m}$ such that $|S \cap f \cdot A^m| \geq (\delta_s + \rho^2)|A|^m$, or
(2) There exists $f \in [A]\binom{n}{m}$ such that

> *(a) $|\{x \in A^m \mid f \cdot x \subset S\}| \geq (\delta_s - 2\delta\rho) \cdot |A|^m$, and*
> *(b) $|\{\ell \in [B]\binom{m}{1} \mid f \cdot \ell \cdot B \subseteq S\}| \geq \delta^{-1}\rho \cdot |[B]\binom{m}{1}|$, where $B \subseteq A$, $|B| = k$.*

If (1) occurs we are happy. We may thus assume that (2) is the case. Our task then is to bring the $(k + 1)$st letter, a_k, back into the game. To accomplish this we introduce the notion of an i-insensitive set: a set $X \subseteq A^n$ is i-insensitive if it has the property that if $x \in X$ is a word and x' is derived from x by replacing some a_i's by a_k's and vice versa, then x' also belongs to X, cf. Sect. 18.3.2 for a formal definition.

Our second step will then be the proof of the following lemma that essentially says that if case (2) occurs in the previous lemma then we can find i-insensitive sets $D_0, \ldots, D_{k-1}$ so that the density of S in $D_0 \cap \ldots \cap D_{k-1}$ is increased.

Lemma 18.7. *Let $k \geq 2$ and assume Theorem 18.1 is true for alphabets of size k. Then there exists for every $\delta > 0$ and $0 < \rho \leq \frac{1}{2}\delta$ an $N_2 = N_2(k + 1, \delta, \rho)$ and $\eta = \eta(k + 1, \delta, \rho) > 0$ such that the following is true. If $|A| = k + 1$, $n \geq N_2$, and $S \subseteq A^n$ satisfies $\delta_s := |S|/|A|^n > \delta$ and*

$$|\{\ell \in [B]\binom{n}{1} \mid \ell \cdot B \subseteq S\}| \geq \tfrac{1}{2}\delta^{-1}\rho \cdot |[B]\binom{n}{1}|,$$

then at least one of the following two properties hold:

(1) There exist $\ell \in [A]\binom{n}{1}$ such that $\ell \cdot A \subseteq S$, or
(2) There exists $D_0, \ldots, D_{k-1} \subseteq A^n$ such that D_i is i-insensitive and such that $D := D_0 \cap \ldots \cap D_{k-1}$ satisfies $|D| \geq \eta |A|^n$ and $|S \cap D| \geq (\delta_s + \frac{1}{3}\rho)|D|$.

In a third step we then use the induction assumption again to show that we can cover the intersection of i-insensitive sets by a collection of pairwise disjoint subspaces.

Lemma 18.8. *Let* $k \geq 2$. *Then there exists for every* $m \in \mathbb{N}$ *and every* $\epsilon > 0$ *an* $N_3 = N_3(k+1, m, \epsilon)$ *such that the following is true. If* $n \geq N_3$ *and* $D_0, \ldots, D_{k-1} \subseteq A^n$ *are sets such that* D_i *is* i-insensitive, *for* $0 \leq i < k$, *and such that* $D := D_0 \cap \ldots \cap D_{k-1}$ *satisfies* $|D| > \epsilon|A|^n$, *then there exists* $f_0, \ldots, f_{v-1} \in [A]^m \binom{n}{m}$ *such that the sets* $X_i := f_i \cdot A^m$ *are pairwise disjoint and satisfy*

$$X_i \subseteq D \qquad \text{for all } 0 \leq i < v$$

and

$$|D \setminus \bigcup_{0 \leq i < v} X_i| \leq \epsilon|A^n|.$$

With Lemmas 18.6–18.8 at hand the proof of the key lemma is easily completed.

Proof of Lemma 18.5. Let $\rho := \rho(k + 1, \delta)$ and $m_0 = m_0(k + 1, \delta)$ where $\rho(\cdot)$ and $m_0(\cdot)$ are as defined in Lemma 18.6. Furthermore, let $\eta := \eta(k+1, \delta-2\delta\rho, \rho)$, where $\eta(\cdot)$ is the function from Lemma 18.7 and set $\epsilon := \frac{1}{12}\rho\eta$ and $\sigma := \min\{\rho^2, \frac{1}{12}\rho\}$. Finally, let

$$m_1 := \max\{N_2(k + 1, \delta - 2\delta\rho, \rho), N_3(k + 1, m, \epsilon), m\},$$

where $N_2(\cdot)$ its the function from Lemma 18.7 and $N_3(\cdot)$ its the function from Lemma 18.8 and let $n_0 := N_1(k + 1, m_1, \delta)$, where $N_1(\cdot)$ its the function from Lemma 18.6.

Assume $n \geq n_0$ and assume without loss of generality that $\delta < \frac{1}{12}$ and $S \subseteq A^n$ satisfies $\delta_s := |S|/|A|^n \geq \delta$. We apply Lemma 18.6 for "m" $= m_1$. We either find $f \in [A]\binom{n}{m_1}$ such that $|S \cap f \cdot A^{m_1}|/|A|^{m_1} \geq \delta_s + \rho^2$. This easily also gives an $f' \in [A]\binom{n}{m}$ such that $|S \cap f' \cdot A^m|/|A|^m \geq \delta_s + \rho^2$, so the proof is completed in this case. Otherwise (2) occurs. Let $S' := \{x \in A^{m_1} \mid f \cdot x \in S\}$ and apply Lemma 18.7 for "n" $= m_1$, S', "δ"$= \delta - 2\rho\delta$ and ρ as set above. Observe that for $\rho < \frac{1}{8}\delta < 1/4$ we have that $(\delta - 2\rho\delta) \geq \frac{1}{2}\delta$ and the assumptions of Lemma 18.7 are thus implied by (2a) and (2b) of Lemma 18.6. If (1) occurs in Lemma 18.7, then $f \cdot \ell$ is the desired line and we are done. So assume (2) occurs. We thus obtain a set $D \subseteq A^{m_1}$ that is the intersection of i-insensitive sets D_i such that $|D| \geq \eta|A|^{m_1}$ and $|S' \cap D|/|D| \geq \delta_s - 2\rho\delta + \frac{1}{3}\rho \geq \delta_s + \frac{1}{6}\rho$, as $\delta < \frac{1}{12}$. We can then apply Lemma 18.8 to cover D with m-dimensional subspaces $f_0, \ldots, f_{v-1}$ such that the uncovered part contains at most $\epsilon|A|^{m_1}$ points. Then

$$\frac{\sum_{0 \leq i \leq v} |f_i \cdot A^m \cap S'|}{\sum_{0 \leq i \leq v} |f_i \cdot A^m|} \geq \frac{|S' \cap D| - \epsilon|A|^{m_1}}{|D|} \geq [(\delta_s + \frac{1}{6}\rho)] - \epsilon\eta^{-1}.$$

Thus, there exists $0 \leq i_0 < \nu$ such that

$$\frac{|f_{i_0} \cdot A^m \cap S'|}{|A|^m} \geq (\delta_s + \tfrac{1}{6}\rho) - \epsilon\eta^{-1} \geq \delta_s + \sigma,$$

by choice of ϵ and σ. Thus $f \cdot f_{i_0}$ is the desired subspace.   □

In the remainder of this section we devote a subsection to the proofs of each of the missing three lemmas (Lemmas 18.6–18.8).

18.3.1   Proof of Lemma 18.6

The basis of the proof of Lemma 18.6 is the following trivial observation that says that whenever one element of a sequence is significantly *below* the average this has to be compensated by at least one element that is a certain value *above* the average:

> Assume $0 \leq t_1, \ldots, t_n \leq 1$ are such that $\frac{1}{n}\sum_i t_i = \mu$. Then   (18.1)
>
> $\exists i$ s.t. $t_i \leq \mu - \epsilon$ $\implies$ $\exists i'$ s.t. $t_{i'} \geq \mu + \frac{1}{n-1}\epsilon.$

This observation has the following consequence: for every $S \subseteq A^n$ with $\delta_s := |S|/|A|^n$ and every $m < n$ we either have that *all* elements in A^m have about the right number of continuations in S or there exists at least one element in A^m that has significantly more continuations in S than expected. Formally:

Lemma 18.9. *For every $k, m \in \mathbb{N}$ and every $\epsilon > 0$ the following is true. If $|A| = k + 1$, $n > m$ and $S \subseteq A^n$ satisfies $\delta_s := |S|/|A|^n > 0$ then at least one of the following two properties hold:*

(1) $|S_x| \geq (\delta_s - \epsilon)|A|^{n-m}$ *for all $x \in A^m$, or*
(2) *There exists x_0 in A^m such that $|S_{x_0}| \geq (\delta_s + \frac{\epsilon}{k^m})|A|^{n-m}$,*

where $S_x := \{z \in A^{n-m} \mid x \times z \in S\}$ for $x \in A^m$

Proof. This follows immediately from (18.1) by considering $t_x := |S_x|/|A|^{n-m}$.

□

With this lemma at hand we can use a density increasing argument in order to find a subspace $f \in [A]\binom{t}{m}$ such that *every* element in this subspace has about the expected number of continuations in A^{n-t} that belong to S.

Corollary 18.10. *For every $k, m \in \mathbb{N}$ and every $\epsilon > 0$ there exists $t_0 = t_0(k + 1, m, \epsilon)$ such that the following is true. If $|A| = k + 1$, $n > t_0$, and $S \subseteq A^n$ satisfies $\delta_s := |S|/|A|^n > 0$ then there exists $t \leq t_0$ and $f \in [A]\binom{t}{m}$ such that $|S_x| \geq (\delta_s - \epsilon)|A|^{n-\ell}$ for all $x \in A^m$, where $S_x := \{z \in A^{n-\ell} \mid f \cdot x \times z \in S\}$.*

Proof. Let $\eta := \frac{\epsilon}{km}$. By Lemma 18.9 we get that either $t = m$ and $f = (\lambda_0, \ldots, \lambda_{m-1})$ has the desired property or there exists $x_1 \in A^m$ such that

$$S_1 := \{z \in A^{n-m} \mid x_1 \times z \in S\} \quad \text{satisfies } |S_1|/|A|^{n-m} \geq \delta_s + \eta.$$

Applying Lemma 18.9 again we get that either $t = 2m$ and $f = x_1 \times (\lambda_0, \ldots, \lambda_{m-1})$ has the desired property or there exists $x_2 \in A^m$ such that

$$S_2 := \{z \in A^{n-2m} \mid x_2 \times z \in S_1\} \quad \text{satisfies } |S_2|/|A|^{n-2m} \geq \delta_s + 2\eta.$$

We now proceed in this way. Clearly, this process can run for at most $\lceil \eta^{-1} \rceil$ steps as $\delta_s + \lceil \eta^{-1} \rceil \cdot \eta > 1$, which can't be. We can thus let $t_0 := \lceil \eta^{-1} \rceil \cdot m$. □

With Corollary 18.10 we have reduced our problem to the case $S \subseteq A^{m+n}$ where *every* element in A^m has lots of continuations in A^n that belong to S. Our next step is to essentially turn this around: we want to argue that this implies that there exists a subset $W \subseteq A^n$ so that *every* element in W has lots of continuations in A^m that belong to S. This will be a consequence of the following observation that is easily proved by contradiction:

Assume $S \subset X \times Y$. Let   (18.2)

$$S_x := \{y \in Y \mid x \times y \in S\} \quad \text{for } x \in X \text{ and}$$

$$T_y := \{x \in X \mid x \times y \in S\} \quad \text{for } y \in Y.$$

Then $|S_x| \geq \mu|Y|$ for all $x \in X$ implies

$$\exists Y_0 \subseteq Y \ : \ |Y_0| \geq \tfrac{1}{2}\mu|Y| \quad \text{and} \quad |T_y| \geq \tfrac{1}{2}\mu|X| \quad \text{for all } y \in Y_0.$$

Actually, we will use observation (18.2) to bring the induction hypothesis into play. For that we ignore the $(k+1)$st letter a_k in the first part of the product space $A^m \times A^n$ for now and just consider the subspace $B^m \times A^n$.

Lemma 18.11. *Let $k \geq 2$ and assume Theorem 18.1 is true for alphabets of size k. Then for every $\epsilon > 0$ and there exists $m_0 = m_0(k+1, \epsilon)$ and $\eta = \eta(k+1, \epsilon)$ such that the following is true. If $|A| = k+1$ and $B \subseteq A$ with $|B| = k$, $m \geq m_0$, $n \geq 1$, and $S \subseteq B^m \times A^n$ is such that*

$$|\{z \in A^n \mid x \times z \in S\}| \geq \epsilon|A|^n \quad \text{for all } x \in B^m,$$

then there exists $\ell \in [B]\binom{m}{1}$ such that

$$|\{z \in A^n \mid \ell \cdot b \times z \in S \text{ for all } b \in B\}| \geq \eta|A|^n. \tag{18.3}$$

Proof. We let $m_0 := DHJ(k, 1, \frac{1}{2}\epsilon)$, where the function $DHJ(\cdot)$ is the one provided by Theorem 18.1. Fix $f \in [B]\binom{m}{m_0}$ arbitrarily. In the following we only consider the m_0-dimensional subspace of B^m induced by f.

By (18.2) we know that there exists $W \subseteq A^n$ of size $|W| \geq \frac{1}{2}\epsilon|A|^n$ such that for every $w \in W$ we have:

$$|\{x \in B^{m_0} \mid f \cdot x \times w \in S\}| \geq \frac{1}{2}\epsilon|B|^{m_0}.$$

By the definition of m_0 this implies that for every $w \in W$ we find a combinatorial line $\ell_w \in [B]\binom{m_0}{1}$ such that

$$f \cdot \ell_w \cdot b \times w \in S \qquad \text{for all } b \in B.$$

Recall that the number of lines in B^{m_0} is $(|B| + 1)^{m_0} - |B|^{m_0} \leq (k + 1)^m$. Hence, there exists $W' \subseteq W$ of size $|W'| \geq \frac{1}{(k+1)^{m_0}}|W| \geq \frac{\epsilon}{2(k+1)^{m_0}}|A^n|$ such that for all $w \in W'$ the lines ℓ_w are identical. In other words, there exists $\ell \in [B]\binom{m_0}{1}$ so that $\ell_w = \ell$ for all $w \in W'$ and thus

$$f \cdot \ell \cdot B \times W' = \{f \cdot \ell \cdot b \times w \mid b \in B, w \in W'\} \subseteq S.$$

For $\eta = \frac{1}{2}\epsilon/(k + 1)^{m_0}$ the line $f \cdot \ell \in [B]\binom{m}{1}$ satisfies (18.3). (Note that the sole purpose of the subspace f is to reduce the dimension from m to m_0; this allowed us to bound the number of lines (and thus η) by a function that only depends on k and ϵ.)   □

By combining this lemma with the Graham-Rothschild Theorem 5.1 we can find a subspace in which *every* line satisfies (18.3):

Corollary 18.12. *Let $k \geq 2$ and assume Theorem 18.1 is true for alphabets of size k. Then for every $\epsilon > 0$ and there exists $m_0 = m_0(k+1, \epsilon)$ and $\eta = \eta(k+1, \epsilon)$ such that such that for every $m \geq m_0$ there exists $M = M(m)$ such that the following is true for all $n \geq 1$. If $|A| = k + 1$ and $B \subseteq A$ with $|B| = k$ and $S \subseteq B^M \times A^n$ such that*

$$|\{z \in A^n \mid x \times z \in S\}| \geq \epsilon|A|^n \qquad \text{for all } x \in B^M,$$

then there exists $f \in [B]\binom{M}{m}$ such that

$$|\{z \in A^n \mid f \cdot \ell \cdot B \times z \subseteq S\}| \geq \eta|A|^n \qquad \text{for all } \ell \in [B]\binom{m}{1}.$$

Proof. Define m_0 and η as in Lemma 18.11. Furthermore, let $M = GR(m, 2)$, where the function $GR(\cdot)$ is the one provided by the Graham-Rothschild Theorem 5.1. Consider the coloring $\Delta : [B]\binom{M}{1}$ defined as follows

$$\Delta(\ell) = \begin{cases} 1 & \text{if } \ell \text{ satisfies equation (18.3)} \\ 0 & \text{otherwise.} \end{cases}$$

Then the Graham-Rothschild Theorem 5.1 implies that there exists an m-dimensional subspace $f \in [A]\binom{M}{m}$ that is monochromatic. Observe that this space cannot be monochromatic in color 0, as by Lemma 18.11 every m-dimensional subspace contains at least one line that satisfies equation (18.3). Hence, f is the desired subspace.   □

With these facts at hand we can now prove Lemma 18.6.

Proof of Lemma 18.6. Let

$$\eta = \eta(k + 1, \tfrac{1}{2}\delta), \qquad m_0 = m_0(k + 1, \tfrac{1}{2}\delta) \qquad \text{and} \qquad M = M(m),$$

where $\eta(\cdot)$, $m_0(\cdot)$, and $M(\cdot)$ are the functions given by Corollary 18.12. Let

$$\rho = \rho(k + 1, \delta) := \tfrac{1}{8}\delta\eta \qquad \text{and} \qquad \epsilon := \tfrac{1}{5}\eta^2\rho^2.$$

Finally, let $T := t_0(k+1, M, \epsilon)$, where $t_0(\cdot)$ is the function given by Corollary 18.10 and set $n_0 := \max\{T + 1, M + 1\}$.

In the following we assume that (1) does not hold, i.e., that every $f \in [A]\binom{n}{m}$ satisfies $|\{x \in A^m \mid f \cdot x \in S\}| \leq (\delta_s + \rho^2)|A|^m$.

Apply Corollary 18.10 for ϵ as defined above and "m" $= M$. This gives us $t \leq T$ and an M-dimensional subspace $f_0 \in [A]\binom{t}{M}$ such that for $S' := \{x \times z \mid x \in A^M, z \in A^{n-t}, f_0 \cdot x \times z \in S\}$ we have

$$|\{z \in A^{n-t} \mid x \times z \in S'\}| \geq (\delta_s - \epsilon)|A|^{n-t} \quad \text{for all } x \in A^M. \tag{18.4}$$

As $\delta_s - \epsilon \geq \tfrac{1}{2}\delta$, we can thus apply Corollary 18.12 with "ϵ"$= \tfrac{1}{2}\delta$ to obtain $f_1 \in [B]\binom{M}{m}$ so that

$$|\{z \in A^{n-t} \mid f_1 \cdot \ell \cdot B \times z \subseteq S'\}| \geq \eta|A|^{n-t} \quad \text{for all } \ell \in [B]\binom{m}{1}. \tag{18.5}$$

Observe that every $z \in A^{n-t}$ gives rise to an m-dimensional subspace $g_z := f_0 \cdot f_1 \times z \in [A]\binom{n}{m}$. Let

$$r_z := \frac{|\{x \in A^m \mid g_z \cdot x \in S\}|}{|A|^m} \qquad \text{for all } z \in A^{n-t}.$$

By our assumption that (1) does not hold, we know that

$$r_z \leq \delta_s + \rho^2 \qquad \text{for all } z \in A^{n-t}.$$

Observe that (18.4) implies that

$$\sum_{z \in A^{n-t}} r_z = \frac{\sum_{x \in A^m} |\{z \in A^{n-t} \mid f_0 \cdot f_1 \cdot x \times z \in S\}|}{|A|^m} \geq (\delta_s - \epsilon)|A|^{n-t}.$$

From these two inequalities we easily deduce that

$$W_1 := \{z \in A^{n-t} \mid r_z \geq \delta_s - 2\delta\rho\}.$$

satisfies $|W_1| \geq (1 - \frac{1}{4}\eta)|A|^{n-t}$. Note that every $z \in W_1$ gives rise to a subspace g_z that satisfies (2a) of the lemma. In order to find a z that also satisfies (2b) let

$$\hat{r}_z := \frac{|\{\ell \in [B]\binom{m}{1} \mid g_z \cdot \ell \cdot B \subseteq S\}|}{|[B]\binom{m}{1}|} \qquad \text{for all } z \in A^{n-t}.$$

Then (18.5) implies that

$$\sum_{z \in A^{n-t}} \hat{r}_z = \frac{\sum_{\ell \in [B]\binom{m}{1}} |\{z \in A^{n-t} \mid f_0 \cdot f_1 \cdot \ell \cdot B \times z \subseteq S\}|}{|[B]\binom{m}{1}|} \geq \eta|A|^{n-t}.$$

Letting

$$W_2 := \{z \in A^{n-t} \mid \hat{r}_z \geq \tfrac{1}{2}\eta = \delta^{-1}\rho\}.$$

satisfies $|W_2| \geq \frac{1}{2}\eta|A|^{n-t}$. One easily checks that $W_1 \cap W_2 \neq \emptyset$ and that any g_w for $w \in W_1 \cap W_2$ satisfies (2a) and (2b).   □

18.3.2   Proof of Lemma 18.7

Our task in the second step is to bring the letter a_k back into the game. The natural approach is to show that a_k is interchangeable with some other letter. To formalize this idea we introduce the following notation. For $x \in A^m$ we denote by $n_{x,i}$ the total number of occurrences of the letters a_i and a_k. Then we let $\xi_{x,i}$ denote the $n_{x,i}$-parameter word that is obtained from x by replacing every occurrence of the letters a_i and a_k by a parameter. For example, if $k = 5$ and $x = (a_4, a_1, a_2, a_1, a_0, a_4)$ then $\xi_{x,0}$ is the 3-parameter word $(\lambda_0, a_1, a_2, a_1, \lambda_1, \lambda_2)$ while $\xi_{x,3}$ is a 2-parameter word: $\xi_{x,3} = (\lambda_0, a_1, a_2, a_1, a_0, \lambda_1)$.

With this notation at hand we are ready to define what we mean by an i-insensitive set. We call a set $W \subset A^n$ i-insensitive if the following is true:

$$x \in W \quad \implies \quad \xi_{x,i} \cdot y \in W \quad \text{for all } y \in \{a_i, a_k\}^{n_{x,i}}.$$

In other words, we require that we can arbitrarily replace some a_k's by a_i's and vice versa without falling out of the set W.

Proof of Lemma 18.7. Choose $N_2 = N_2(k+1, \rho)$ such that $[k/(k+1)]^n \leq \frac{1}{9}\delta^{-1}\rho^2$, $\eta = \eta(k+1, \delta, \rho) = \frac{\rho}{6k}(\delta - \frac{1}{9}\delta^{-1}\rho^2)$ and let

$$\mathcal{L} := \{\ell \in [B]\binom{n}{1} \mid \ell \cdot B \subseteq S\} \quad \text{and} \quad L := \{\ell \cdot a_k \mid \ell \in \mathcal{L}\}.$$

If $S \cap L \neq \emptyset$ then $A^n \cap S$ contains a combinatorial line and we are done. So assume this is not the case. For $x \in A^n \setminus B^n$ let $\ell_x \in [B]\binom{n}{1}$ denote the line given by x in which every occurrence of the $(k + 1)$st letter (i.e., every occurrence of a_k) is replaced by the parameter λ_0. Let

$$X_i := \{x \in A^n \setminus B^n \mid \ell_x \cdot a_i \in S\} \cup (B^n \cap S).$$

Clearly, for $x \in A^n \setminus B^n$ we have

$$x \in \bigcap_{0 \leq i < k} X_i \quad \Longleftrightarrow \quad \ell_x \in \mathcal{L}. \tag{18.6}$$

Observe also that the X_i are i-insensitive. (To see this consider $x \in A^n$ and assume x' is obtained from x by replacing some a_k's by a_i's. Then $\ell_x \cdot a_i = \ell_{x'} \cdot a_i$. That is, either none of x and x' belongs to the set X_i or both do.) Finally, let

$$X = X_0 \cap \ldots \cap X_{k-1}$$

and observe that (18.6) implies $L \subseteq X$ and thus

$$|X| \geq |L| = |\mathcal{L}| \geq \frac{1}{2}\delta^{-1}\rho |[B]\binom{n}{1}| = \frac{1}{2}\delta^{-1}\rho \cdot [(k + 1)^n - k^n] \geq \frac{1}{3}\delta^{-1}\rho |A|^n,$$

by choice of n_0, with room to spare. Furthermore, from (18.6) we know that $X \subseteq L \cup B^n$ and the assumption $S \cap L = \emptyset$ thus implies

$$|S \cap X| \leq |B^n| = (\tfrac{k}{k+1})^n |A|^n \leq \tfrac{1}{9}\delta^{-1}\rho^2 |A|^n, \tag{18.7}$$

by choice of n_0. Combing both inequalities we get

$$\frac{|S \cap (A^n \setminus X)|}{|A^n \setminus X|} \geq \frac{\delta_s - \frac{1}{9}\delta^{-1}\rho^2}{1 - \frac{1}{3}\delta^{-1}\rho} \geq \delta_s + \tfrac{1}{3}\rho.$$

We define a partition of $A^n \setminus X$ as follows. $P_0 = A^n \setminus X_0$ and $P_i = (A^n \setminus X_i) \cap (X_0 \cap \ldots \cap X_{i-1})$ for $0 < i < k$. Let $I := \{0 \leq i < k \mid |P_i| \geq \frac{\rho}{6k}|A^n \setminus X|\}$. Then

$$\delta_s + \tfrac{1}{3}\rho \leq \frac{|S \cap (A^n \setminus X)|}{|A^n \setminus X|} \leq \frac{\sum_{i \in I} |S \cap P_i| + \tfrac{1}{6}\rho|A^n \setminus X|}{\sum_i |P_i|} \leq \frac{\sum_{i \in I} |S \cap P_i|}{\sum_{i \in I} |P_i|} + \tfrac{1}{6}\rho$$

and we deduce that there has to exist $i_0 \in I$ such that $|S \cap P_{i_0}|/|P_{i_0}| \geq \delta_s + \tfrac{1}{6}\rho$. (Here we used that if $a_i/b_i < x$ for all i, then $(\sum a_i)/(\sum b_i) < x$.) By definition of I we have $|P_{i_0}| \geq \tfrac{\rho}{6k}|A^n \setminus X| \geq \tfrac{\rho}{6k}|S \cap (A^n \setminus X)| \geq \eta|A|^n$, by definition of η and (18.7). Let

$$D_i = \begin{cases} X_i & \text{if } i < i_0 \\ A^n \setminus X_{i_0} & \text{if } i = i_0 \\ A^n & \text{if } i > i_0. \end{cases}$$

Clearly, all D_i are i-insensitive. Moreover $\bigcap D_i = P_{i_0}$ and thus densities are as required.   □

18.3.3   Proof of Lemma 18.8

Lemma 18.8 claims that we can cover the intersection of k insensitive sets by a collection of pairwise disjoint subspaces. The proof of this lemma comes in three parts. First we show that if $S \subseteq A^n$ satisfies $|S|/|A|^n \geq \delta$ then we can find an m-dimensional subspace f such that $f \cdot B^m$ is contained in S. Then we show that this result implies that a result similar to Lemma 18.8 holds if we consider instead of the intersection of k insensitive sets just a single insensitive set. In the final part of this section we then prove the Lemma 18.8 by an induction on the number of insensitive sets that form the intersection.

Lemma 18.13. *Let $k \geq 2$ and assume Theorem 18.1 is true for alphabets of size k. Then there exists for every integer $m \in \mathbb{N}$ and every $\epsilon > 0$ an $n_0 = n_0(k+1, m, \epsilon)$ such that the following is true. If $|A| = k+1$ and $B \subseteq A$ with $|B| = k$, $n \geq n_0$, and $S \subseteq A^n$ satisfies $|S| \geq \epsilon|A^n|$, then there exists $f \in [A]\binom{n}{m}$ such that $f \cdot B \subseteq S$.*

Proof. Let $M = DHJ(k, m, \epsilon/2)$ and let $T := t_0(k+1, M, \epsilon/2)$, where t_0 is the function $t_0(\cdot)$ from Corollary 18.10. Let $n_0 = T + 1$.

By Corollary 18.10 we know that there exists $t \leq T$ and $f \in [A]\binom{t}{M}$ such that $S_x := \{z \in A^{n-t} \mid f \cdot x \times z \in S\}$ satisfies $|S_x| \geq (|S|/|A|^n - \tfrac{1}{2}\epsilon)|A|^{n-t} \geq \tfrac{1}{2}\epsilon|A|^{n-t}$ for every $x \in A^M$. Hence,

$$\sum_{x \in B^M} |S_x| \geq \tfrac{1}{2}\epsilon|A|^{n-t} \cdot |B|^M$$

and there thus has to exist $z_0 \in A^{n-t}$ that is contained in at least $\tfrac{1}{2}\epsilon|B|^M$ many S_x with $x \in B^M$. Consider $f' := f \times z_0 \in [A]\binom{n}{M}$ and let $S' := \{x \in B^M \mid f' \cdot x \in S\}$.

By choice of z_0 we know $|S'| \geq \frac{1}{2}\epsilon|B|^m$ and we can thus apply Theorem 18.1 to S' and B^M to find a $g \in [B]\binom{M}{m}$ such that $g \cdot B^m \subseteq S'$. Then $f' \cdot g \in [A]\binom{n}{m}$ is the desired space.   □

Lemma 18.14. *Let $k \geq 2$ and assume Theorem 18.1 is true for alphabets of size k. Then there exists for every integer $m \in \mathbb{N}$ and every $\epsilon > 0$ an $n = n(k+1, m, \epsilon)$ such that the following is true. If $|A| = k+1$ and $B \subseteq A$ with $|B| = k$, $n \geq n_0$ and $D \subseteq A^n$ is i-insensitive, for some $0 \leq i < k$, with $|D| > \epsilon|A^n|$, then there exists $f_0, \dots, f_{\nu-1} \in [A]\binom{n}{m}$ such that the sets $X_j := f_j \cdot A^m$ are pairwise disjoint and satisfy*

$$X_j \subseteq D \quad \text{for all } 0 \leq j < \nu \qquad \text{and} \qquad \left|D \setminus \bigcup_{0 \leq j < \nu} X_j\right| \leq \epsilon|A^n|.$$

Proof. Let $M = n(k+1, m, \frac{1}{2}\epsilon)$, where $n(\cdot)$ is the function provided by Lemma 18.13, and let $\tau := \epsilon/(2(k+m)^{2M})$. We construct the subspaces f_j in at most $\lceil \tau^{-1} \rceil$ rounds. In the jth round we construct a parameter word $g \in [A]\binom{M}{m}$ in the M-dimensional subspace given by the coordinates $(j-1)M, \dots, jM-1$. We then extend g by appropriate words $\xi \in A^{(j-1)M}$ and $z \in A^{n-jM}$ to a word $f_{\xi,z} = \xi \times g \times z \in [A]\binom{n}{m}$ to obtain a collection of subspaces that are contained in D and are pairwise disjoint from all previously constructed subsets. We will guarantee that in the jth round we find at least $\tau|A^n|$ such subspaces. Thus $\lceil \tau^{-1} \rceil$ rounds will certainly suffice to cover D. To formalize this idea we prove the following:

(∗)   Assume $0 \leq j < \lceil \tau^{-1} \rceil$ and assume that for every $\xi \in A^{jM}$ there exists an i-insensitive set $D_\xi \subseteq A^{n-jM}$ such that $\sum_\xi |D_\xi| > \epsilon|A^n|$. Then there exists $g \in [A]\binom{M}{m}$ and sets $Z_\xi \subseteq A^{n-(j+1)M}$ such that $g \cdot A^m \times z \in D_\xi$ for all $z \in Z_\xi$ and all $\xi \in A^{jM}$ and all sets Z_ξ are i-insensitive and satisfy $\sum_\xi |Z_\xi| \geq \tau|A^n|$.

This can be seen as follows. We partition $A^{jM} \times A^M \times A^{n-(j+1)M}$ into subsets by fixing the first and the last part. That is, for every pair $\xi \in A^{jM}$ and $z \in A^{n-(j+1)M}$ we let

$$D_{\xi,z} := \{y \in A^M \mid y \times z \in D_\xi\}.$$

Using the idea from (18.2) we deduce that there exists $W \subseteq A^{jM} \times A^{n-(j+1)M}$ such that

$$|W| \geq \frac{1}{2}\epsilon|A|^{n-M} \qquad \text{and} \qquad |D_{\xi,z}| \geq \frac{1}{2}\epsilon|A|^M \quad \text{for all } (\xi, z) \in W.$$

Lemma 18.13 then implies that we find for every pair $(\xi, z) \in W$ an $f_{\xi,z} \in [B]\binom{M}{m}$ such that

$$f_{\xi,z} \cdot B^m \subseteq D_{\xi,z}.$$

By assumption D_ξ is i-insensitive. Note that the definition of the sets $D_{\xi,z}$ ensures that these sets inherit the i-insensitivity from D_ξ. Thus we actually have

$$f_{\xi,z} \cdot A^m \subseteq D_{\xi,z} \quad \text{and thus also} \quad f_{\xi,z} \cdot A^m \times z \subseteq D_\xi.$$

Recall that $|[B]\binom{M}{m}| \leq (k+m)^M$. We thus deduce that there exists a $g \in [B]\binom{M}{m}$ so that

$$|\{(\xi,z) \in W \mid f_{\xi,z} = g\}| \geq \tfrac{1}{(k+m)^M}|W| \geq \tfrac{\epsilon}{2(k+m)^M}|A|^{n-M} \geq \tau|A|^n.$$

For every $\xi \in A^{jM}$ let

$$Z_\xi := \{z \in A^{n-(j+1)M} \mid g \cdot A^m \times z \subseteq D_\xi\}.$$

Clearly, the sets Z_ξ satisfy the properties claimed in (∗).

Now we show how to apply (∗) in order to prove the lemma. The idea is to define the sets D_ξ in such a way that they together cover exactly those words from D that are not contained in any of the subspace constructed so far.

For $j = 0$ the assumptions are obviously satisfied. Assume we applied (∗) for some $j \geq 1$. We need to explain how to setup the sets D_ξ for the next round. Consider $\xi' = \xi \times y \in A^{jM} \times A^M$. We set

$$D_{\xi'} = \begin{cases} \{z \in A^{n-(j+1)M} \mid y \times z \in D_\xi\} & \text{if } y \notin g \cdot A^m \\ \{z \in A^{n-(j+1)M} \mid y \times z \in D_\xi\} \setminus Z_\xi & \text{otherwise.} \end{cases}$$

For $y \notin g \cdot A^m$ one easily checks that the set $D_{\xi'}$ inherits the i-insensitivity from D_ξ. For $y \in g \cdot A^m$ this follows similarly, as Z_ξ is i-insensitive and the difference of two i-insensitive sets is again i-insensitive. Observe also that

$$\bigcup_{\xi' \in A^{(j+1)M}} \xi' \times D_{\xi'} \cup \bigcup_{\xi \in A^{jM}} \bigcup_{z \in Z_\xi} \xi \times g \cdot A^m \times z = \bigcup_{\xi \in A^{jM}} \xi \times D_\xi.$$

That is, the new sets $D_{\xi'}$ cover exactly those words from D that are not contained in any of the subspaces constructed so far.   □

Proof of Lemma 18.8. We make repeated use of the previous lemma. Define a sequence $m_r, \ldots, m_0$ as follows. Let

$$m_k := m \quad \text{and} \quad m_{i-1} := n(k+1, m_i, \rho, \tfrac{\epsilon}{k}) \quad \text{for } i = k, \ldots, 1,$$

where the function $n(\cdot)$ is as defined in Lemma 18.14. We claim that the following is true for all $r = 0, \ldots, k-1$:

(∗) There exists $f_0^{(r)}, \ldots, f_{v_r}^{(r)} \in [A]\binom{n}{m_{r+1}}$ such that the sets $X_j := f_j \cdot A^{m_{r+1}}$ are pairwise disjoint and satisfy

$$X_j \subseteq \bigcap_{0 \leq i \leq r} D_i \quad \text{for all } 0 \leq j \leq \nu_r$$

and

$$\Big| \bigcap_{0 \leq i \leq r} D_i \ \setminus \ \bigcup_{0 \leq j \leq \nu_r} X_j \Big| \leq \frac{r+1}{k} \epsilon |A^n|.$$

Clearly, $(*)$ holds for $r = 0$ if we set $n(k, m, \epsilon) := m_0$. So assume we have shown $(*)$ for some $0 \leq r < k - 1$. The inductive assumption gives us $f_0^{(r)}, \ldots, f_{\nu_r}^{(r)} \in [A]\binom{n}{m_{r+1}}$. For each $0 \leq j \leq \nu_r$ consider $Y_j := f_j^{(r)} \cdot A^{m_{r+1}} \cap D_{r+1}$. If $|Y_j| > \frac{\epsilon}{k}|A|^{m_{r+1}}$, we can apply Lemma 18.14 to obtain a collection of subspaces (f_s) in $[A]\binom{m_{r+1}}{m_{r+2}}$ so that the union of the subspaces $f_j^{(r)} \cdot f_s$ cover all but at most $\frac{\epsilon}{k}|A|^{m_{r+1}}$ elements of the set Y_j. As, obviously, $\nu_r \leq |A|^{n-m_{r+1}}$ this implies that the union of all these subspaces $f_j^{(r)} \cdot f_s$ cover all but at most $\frac{\epsilon}{k}|A|^n$ elements of $\bigcup_j Y_j$. The claim follows. $\qquad\qquad\square$

References

Abbott, H.L., Hanson, D.: A problem of Schur and its generalizations. Acta Arith. **20**, 175–187 (1972)

Abramson, F.G., Harrington, L.A.: Models without indiscernibles. J. Symb. Log. **43**, 572–600 (1978)

Ajtai, M., Komlós, J., Szemerédi, E.: A note on Ramsey numbers. J. Comb. Theory A **29**, 354–360 (1980)

Ajtai, M., Komlós, J., Szemerédi, E.: A dense infinite Sidon sequence. Eur. J. Comb. **2**, 1–11 (1981)

Alon, N., Maass, W.: Meanders, Ramsey theory and lower bounds for branching programs. In: 27th Symposium on Foundations of Computer Science, Toronto, pp. 410–417 (1986)

Balogh, J., Morris, R., Samotij, W.: Independent sets in hypergraphs (2012). Preprint, arXiv: 1204.6530

Baumgartner, J.E.: A short proof of Hindman's theorem. J. Comb. Theory A **17**, 384–386 (1974)

Baumgartner, J.E.: Canonical partition relations. J. Symb. Log. **40**, 541–554 (1975)

Beck, J., Sós, V.T.: Discrepancy theory. In: Handbook of Combinatorics, vols. 1, 2, pp. 1405–1446. Elsevier, Amsterdam (1995)

Benzait, A., Voigt, B.: A combinatorial interpretation of $(1/k!)\Delta^k t^n$. Discrete Math. **73**, 27–35 (1989)

Berlekamp, E.R.: A construction for partitions which avoid long arithmetic progressions. Can. Math. Bull. **11**, 409–414 (1968)

Birkhoff, G.: Lattice Theory, 3rd edn. American Mathematical Society, Providence (1967)

Bohman, T.: The triangle-free process. Adv. Math. **221**, 1653–1677 (2009)

Bohman, T., Keevash, P.: The early evolution of the H-free process. Invent. Math. **181**, 291–336 (2010)

Bollobás, B.: Modern Graph Theory. Springer, New York (1998)

Bollobás, B.: Random Graphs, 2nd edn. Cambridge University Press, Cambridge (2001)

Braithwaite, R.: Editor's introduction. In: Ramsey, F.P., Braithwaite, R.B. (ed.) The Foundations of Mathematics. Routledge & Kegan Paul, London (1931)

Brauer, A.: Über Sequenzen von Potenzresten. Sitzungsber. Preuß. Akad. Wiss. Math. Phy. Kl pp. 9–16 (1928)

Brauer, A.: Gedenkrede auf Issai Schur. In: Brauer, A., Rohrbach, H. (eds.) Issai Schur, Gesammelte Abhandlungen, pp. v–xiv. Springer, Berlin (1973)

Brown, T.C., Buhler, J.P.: Lines imply spaces in density Ramsey theory. J. Comb. Theory A **36**, 214–220 (1984)

Brown, T.C., Erdős, P., Chung, F.R.K., Graham, R.L.: Quantitative forms of a theorem of Hilbert. J. Comb. Theory A **38**, 210–216 (1985)

H.J. Prömel, *Ramsey Theory for Discrete Structures*,
DOI 10.1007/978-3-319-01315-2,
© Springer International Publishing Switzerland 2013

Buchholz, W., Wainer, S.: Provably computable functions and the fast growing hierarchy. In: Logic and Combinatorics (Arcata, Calif., 1985), pp. 179–198. American Mathematical Society, Providence (1987)

Cameron, P.J.: Aspects of the random graph. In: Graph Theory and Combinatorics (Cambridge, 1983), pp. 65–79. Academic, London (1984)

Carlitz, L.: Weighted stirling numbers of the first and second kind. I. Fibonacci Quart. **18**, 147–162 (1980)

Chazelle, B.: The Discrepancy Method. Cambridge University Press, Cambridge (2000)

Chen, B., Matsumoto, M., Wang, J.F., Zhang, Z.F., Zhang, J.X.: A short proof of Nash-Williams' theorem for the arboricity of a graph. Graphs Comb. **10**, 27–28 (1994)

Chernoff, H.: A measure of asymptotic efficiency for tests of a hypothesis based on the sum of observations. Ann. Math. Stat. **23**, 493–507 (1952)

Chung, F.R.K.: On the Ramsey numbers $N(3, 3, \ldots, 3; 2)$. Discrete Math. **5**, 317–321 (1973)

Chung, F.R.K., Grinstead, C.M.: A survey of bounds for classical Ramsey numbers. J. Graph Theory **7**, 25–37 (1983)

Chvátal, V.: Some unknown van der Waerden numbers. In: Combinatorial Structures and their Applications Proceedings Calgary International Conference, Calgary, 1969, pp. 31–33. Gordon and Breach, New York (1970)

Conlon, D.: A new upper bound for diagonal Ramsey numbers. Ann. Math. **170**, 941–960 (2009)

Descartes, B.: A three color problem. Eureka, University of Cambridge (1948)

Deuber, W.: Partitionen und lineare Gleichungssysteme. Math. Z. **133**, 109–123 (1973)

Deuber, W.: Partitionstheoreme für Graphen. Comment. Math. Helv. **50**, 311–320 (1975)

Deuber, W.: On van der Waerden's theorem on arithmetic progressions. J. Comb. Theory A **32**, 115–118 (1982)

Deuber, W., Rothschild, B.L., Voigt, B.: Induced partition theorems. J. Comb. Theory A **32**, 225–240 (1982)

Deuber, W., Graham, R.L., Prömel, H.J., Voigt, B.: A canonical partition theorem for equivalence relations on $\mathbf{Z}^t$. J. Comb. Theory A **34**, 331–339 (1983)

Diestel, R.: Graph Theory, 4th edn. Springer, Heidelberg (2010)

Dodos, P., Kanellopoulos, V., Tyros, K.: A simple proof of the density Hales-Jewett theorem. Intern. Math. Res. Not. (2013, to appear). Available in online http://imrn.oxfordjournals.org/content/early/2013/03/15/imrn.rnt041.abstract

Dowling, T.A.: A class of geometric lattices based on finite groups. J. Comb. Theory B **14**, 61–86 (1973)

Dushnik, B., Miller, E.W.: Partially ordered sets. Am. J. Math. **63**, 600–610 (1941)

El-Zahar, M., Sauer, N.: The indivisibility of the homogeneous k_n-free graphs. J. Comb. Theory B **47**, 162–170 (1989)

Erdős, P.: On a lemma of Littlewood and Offord. Bull. Am. Math. Soc. **51**, 898–902 (1945)

Erdős, P.: Some remarks on the theory of graphs. Bull. Am. Math. Soc. **53**, 292–294 (1947)

Erdős, P.: Graph theory and probability. Can. J. Math. **11**, 34–38 (1959)

Erdős, P.: Graph theory and probability. II. Can. J. Math. **13**, 346–352 (1961)

Erdős, P.: On combinatorial questions connected with a theorem of Ramsey and van der Waerden (in hungarian). Mat. Lapok **14**, 29–37 (1963)

Erdős, P.: On the combinatorial problems which I would most like to see solved. Combinatorica **1**, 25–42 (1981)

Erdős, P.: Welcoming address. In: Random Graphs '83 (Poznań, 1983), pp. 1–5. North-Holland, Amsterdam (1985)

Erdős, P., Graham, R.L.: Old and new problems and results in combinatorial number theory. In: Monographies de L'Enseignement Mathématique, vol. 28. L'Enseignement mathématique, Geneva (1980)

Erdős, P., Rado, R.: A combinatorial theorem. J. Lond. Math. Soc. **25**, 249–255 (1950)

Erdős, P., Rado, R.: Combinatorial theorems on classifications of subsets of a given set. Proc. Lond. Math. Soc. **2**, 417–439 (1952)

Erdős, P., Rado, R.: A problem on ordered sets. J. Lond. Math. Soc. **28**, 426–438 (1953)

Erdős, P., Rado, R.: A partition calculus in set theory. Bull. Am. Math. Soc. **62**, 427–489 (1956)

Erdős, P., Rényi, A.: On the evolution of random graphs. Magyar Tud. Akad. Mat. Kutató Int. Közl. **5**, 17–61 (1960)

Erdős, P., Rényi, A.: Asymmetric graphs. Acta Math. Acad. Sci. Hung. **14**, 295–315 (1963)

Erdős, P., Spencer, J.: Imbalances in k-colorations. Networks **1**, 379–385 (1971/1972)

Erdős, P., Szekeres, G.: A combinatorial problem in geometry. Compos. Math. **2**, 463–470 (1935)

Erdős, P., Szekeres, G.: On some extremum problems in elementary geometry. Ann. Univ. Sci. Bp. Eötvös Sect. Math. **3–4**, 53–62 (1961)

Erdős, P., Turán, P.: On some sequences of integers. J. Lond. Math. Soc. **11**, 261–264 (1936)

Erdős, P., Hajnal, A., Rado, R.: Partition relations for cardinal numbers. Acta Math. Acad. Sci. Hung. **16**, 93–196 (1965)

Erdős, P., Hajnal, A., Pósa, L.: Strong embeddings of graphs into colored graphs. In: Infinite And Finite Sets (Colloq., Keszthely, 1973; Dedicated to P. Erdős on his 60th Birthday), vol. I, pp. 585–595. North-Holland, Amsterdam (1975)

Erdős, P., Hajnal, A., Máté, A., Rado, R.: Combinatorial Set Theory: Partition Relations for Cardinals. North-Holland, Amsterdam (1984)

Fettes, S.E., Kramer, R.L., Radziszowski, S.P.: An upper bound of 62 on the classical Ramsey number $R(3, 3, 3, 3)$. Ars Combin. **72**, 41–63 (2004)

Fodor, G.: Eine Bemerkung zur Theorie der regressiven Funktionen. Acta Sci. Math. Szeged **17**, 139–142 (1956)

Folkman, J.: Graphs with monochromatic complete subgraphs in every edge coloring. SIAM J. Appl. Math. **18**, 19–24 (1970)

Frankl, P., Graham, R.L., Rödl, V.: Induced restricted Ramsey theorems for spaces. J. Comb. Theory A **44**, 120–128 (1987)

Fredricksen, H.: Five sum-free sets. In: Proceedings of the Sixth Southeastern Conference on Combinatorics, Graph Theory and Computing (Florida Atlantic University, Boca Raton, 1975). Congressus Numerantium, vol. XIV, pp. 309–314. Utilitas Mathematica, Winnipeg (1975)

Fredricksen, H.: Schur numbers and the Ramsey numbers $N(3, 3, \cdots, 3; 2)$. J. Comb. Theory A **27**, 376–377 (1979)

Furstenberg, H.: Ergodic behavior of diagonal measures and a theorem of Szemerédi on arithmetic progressions. J. Anal. Math. **31**, 204–256 (1977)

Furstenberg, H.: Recurrence in Ergodic Theory and Combinatorial Number Theory. M.B. Porter Lectures. Princeton University Press, Princeton (1981)

Furstenberg, H., Katznelson, Y.: An ergodic Szemerédi theorem for commuting transformations. J. Anal. Math. **34**, 275–291 (1979, 1978)

Furstenberg, H., Katznelson, Y.: A density version of the Hales-Jewett theorem. J. Anal. Math. **57**, 64–119 (1991)

Furstenberg, H., Weiss, B.: Topological dynamics and combinatorial number theory. J. Anal. Math. **34**, 61–85 (1978)

Furstenberg, H., Katznelson, Y., Ornstein, D.: The ergodic theoretical proof of Szemerédi's theorem. Bull. Am. Math. Soc. (N.S.) **7**, 527–552 (1982)

Gödel, K.: Über formal unentscheidbare Sätze der Principia Mathematica und verwandter Systeme. Monatsh. Math. Phys. **38**, 173–198 (1931)

Gowers, W.: A new proof of Szemerédi's theorem. Geom. Funct. Anal. **11**, 465–588 (2001)

Gowers, W.T.: Quasirandomness, counting and regularity for 3-uniform hypergraphs. Comb. Probab. Comput. **15**, 143–184 (2006)

Gowers, W.T.: Hypergraph regularity and the multidimensional Szemerédi theorem. Ann. Math. **166**, 897–946 (2007)

Graham, R.L.: Rudiments of Ramsey Theory. CBMS Regional Conference Series in Mathematics, vol. 45. American Mathematical Society, Providence (1981)

Graham, R.L., Leeb, K., Rothschild, B.L.: Ramsey's theorem for a class of categories. Adv. Math. **8**, 417–433 (1972)

Graham, R.L., Rothschild, B.L.: Ramsey's theorem for n-parameter sets. Trans. Am. Math. Soc. **159**, 257–292 (1971)

Graham, R.L., Rothschild, B.L.: A short proof of van der Waerden's theorem on arithmetic progressions. Proc. Am. Math. Soc. **42**, 385–386 (1974)

Graham, R.L., Spencer, J.H.: A general Ramsey product theorem. Proc. Am. Math. Soc. **73**, 137–139 (1979)

Graham, R.L., Rothschild, B.L., Spencer, J.H.: Ramsey Theory. Wiley, New York (1980)

Grätzer, G.: General Lattice Theory, 2nd edn. Birkhäuser Verlag, Basel (1998)

Graver, J.E., Yackel, J.: Some graph theoretic results associated with Ramsey's theorem. J. Comb. Theory **4**, 125–175 (1968)

Greenwood, R.E., Gleason, A.M.: Combinatorial relations and chromatic graphs. Can. J. Math. **7**, 1–7 (1955)

Hales, A.W., Jewett, R.I.: Regularity and positional games. Trans. Am. Math. Soc. **106**, 222–229 (1963)

Hardy, G.: A theorem concerning the infinite cardinal numbers. Quart J Math **35**, 87–94 (1904)

Heisenberg, W.: Die Abstraktion in der modernen Naturwissenschaft, Talk given at Workshop of "Ordens Pour le mérite für Wissenschaft und Künste" (in German), Bonn 1960. Schritte über Grenzen, Piper, München (1984)

Henson, C.: A family of countable homogeneous graphs. Pac. J. Math. **38**, 69–83 (1971)

Hilbert, D.: Über die Irreducibilität ganzer rationaler Funktionen mit ganzzahligen Coefizienten. J. Reine Angew. Math. **110**, 104–129 (1892)

Hindman, N.: Finite sums from sequences within cells of a partition of N. J. Comb. Theory A **17**, 1–11 (1974)

Hindman, N.: Ultrafilters and combinatorial number theory. In: Number Theory, Carbondale, 1979. Proceedings of the Southern Illinois Number Theory Conference, pp. 119–184. Springer, Berlin (1979)

Irving, R.W.: An extension of Schur's theorem on sum-free partitions. Acta Arith. **25**, 55–64 (1973)

Jech, T.: Set Theory. Pure and Applied Mathematics. Springer, Berlin (1978)

Ježek, J., Nešetřil, J.: Ramsey varieties. Eur. J. Comb. **4**, 143–147 (1983)

Johnson, S.: A new proof of the Erdős-Szekeres convex k-gon result. J. Comb. Theory A **42**, 318–319 (1986)

Kalbfleisch, J.G.: Chromatic graphs and Ramsey's theorem. Phd. Thesis, University of Waterloo (1966)

Ketonen, J., Solovay, R.: Rapidly growing Ramsey functions. Ann. Math. **113**, 267–314 (1981)

Khinchin, A.Y.: Three Pearls of Number Theory. Graylock, Rochester (1952)

Kim, J.H.: The Ramsey number $R(3, t)$ has order of magnitude $t^2/\log t$. Random Struct. Algorithms **7**, 173–207 (1995)

Kleene, S.C.: Introduction to Metamathematics. D. Van Nostrand, New York (1952)

Kleinberg, E.M.: Strong partition properties for infinite cardinals. J. Symb. Log. **35**, 410–428 (1970)

Komjáth, P., Rödl, V.: Coloring of universal graphs. Graphs Comb. **2**, 55–60 (1986)

König, D.: Über eine Schlußweise aus dem Endlichen ins Unendliche. Acta Litt. Acad. Sci. Hung. (Szeged) **3**, 121–130 (1927)

Leeb, K.: A full Ramsey-theorem for the Deuber-category. In: Infinite and Finite Sets (Colloq., Keszthely, 1973; dedicated to P. Erdős on his 60th Birthday), vol. II, pp. 1043–1049. North-Holland, Amsterdam (1975)

Littlewood, J.E., Offord, A.C.: On the number of real roots of a random algebraic equation. III. Rec. Math. [Mat. Sbornik] N.S. **12**, 277–286 (1943)

Loebl, M., Nešetřil, J.: Fast and slow growing (a combinatorial study of unprovability). In: Surveys in Combinatorics, 1991 (Guildford, 1991), pp. 119–160. Cambridge University Press, Cambridge (1991)

Łuczak, T., Ruciński, A., Voigt, B.: Ramsey properties of random graphs. J. Comb. Theory B **56**, 55–68 (1992)

Mathias, A.: On a generalization of Ramsey's theorem. Doctoral dissertation, Cambridge University (1969)

Mellor, D.: The eponymous F.P. Ramsey. J. Graph Theory **7**, 9–13 (1983)

Meyer auf der Heide, F., Wigderson, A.: The complexity of parallel sorting. SIAM J. Comput. **16**, 100–107 (1987)

Moran, S., Snir, M., Manber, U.: Applications of Ramsey's theorem to decision tree complexity. J. Assoc. Comput. Mach. **32**, 938–949 (1985)

Mycielski, J.: Sur le coloriage des graphs. Colloq. Math. **3**, 161–162 (1955)

Nagle, B., Rödl, V., Schacht, M.: The counting lemma for regular k-uniform hypergraphs. Random Struct. Algorithms **28**, 113–179 (2006)

Nenadov, R., Steger, A.: A short proof of the random Ramsey theorem. Comb. Probab. Comput. (to appear)

Nešetřil, J.: Some nonstandard Ramsey-like applications. Theor. Comput. Sci. **34**, 3–15 (1984)

Nešetřil, J., Rödl, V.: Partitions of subgraphs. In: Recent Advances in Graph Theory (Proceedings Second Czechoslovak Symposium, Prague 1974), pp. 413–423. Academia, Prague (1975)

Nešetřil, J., Rödl, V.: Partitions of vertices. Comment. Math. Univ. Carol. **17**, 85–95 (1976a)

Nešetřil, J., Rödl, V.: The Ramsey property for graphs with forbidden complete subgraphs. J. Comb. Theory B **20**, 243–249 (1976b)

Nešetřil, J., Rödl, V.: Partitions of finite relational and set systems. J. Comb. Theory A **22**, 289–312 (1977)

Nešetřil, J., Rödl, V.: On a probabilistic graph-theoretical method. Proc. Am. Math. Soc. **72**, 417–421 (1978a)

Nešetřil, J., Rödl, V.: Selective graphs and hypergraphs. Ann. Discrete Math. **3**, 181–189 (1978b)

Nešetřil, J., Rödl, V.: Ramsey theorem for classes of hypergraphs with forbidden complete subhypergraphs. Czechoslovak Math. J. **29**, 202–218 (1979)

Nešetřil, J., Rödl, V.: Another proof of the Folkman-Rado-Sanders theorem. J. Comb. Theory A **34**, 108–109 (1983a)

Nešetřil, J., Rödl, V.: Ramsey classes of set systems. J. Comb. Theory A **34**, 183–201 (1983b)

Nešetřil, J., Rödl, V.: Two remarks on Ramsey's theorem. Discrete Math. **54**, 339–341 (1985)

Nešetřil, J., Rödl, V. (eds.): Mathematics of Ramsey Theory. Springer, Berlin (1990)

Olivatro, D.: Noughts and crosses. The Sciences Nov/Dec pp. 59–60 (1984)

Paris, J., Harrington, L.: A mathematical incompleteness in Peano arithmetic. In: Barwise, J. (ed.) Handbook of Mathematical Logic, pp. 1133–1142. North Holland, Amsterdam (1977)

Polymath, D.H.J.: A new proof of the density Hales-Jewett theorem. Ann. Math. **175**, 1283–1327 (2012)

Pouzet, M., Rosenberg, I.G.: Ramsey properties for classes of relational systems. Eur. J. Comb. **6**, 361–368 (1985)

Prömel, H.J., Rödl, V.: An elementary proof of the canonizing version of Gallai-Witt's theorem. J. Comb. Theory A **42**, 144–149 (1986)

Prömel, H.J., Rothschild, B.L.: A canonical restricted version of van der Waerden's theorem. Combinatorica **7**, 115–119 (1987)

Prömel, H.J., Voigt, B.: Recent results in partition (Ramsey) theory for finite lattices. Discrete Math. **35**, 185–198 (1981a)

Prömel, H.J., Voigt, B.: Partition theorems for parameter systems and graphs. Discrete Math. **36**, 83–96 (1981b)

Prömel, H.J., Voigt, B.: Canonical partition theorems for finite distributive lattices. In: Proceedings of the 10th Winter School on Abstract Analysis (Srní, 1982), Palermo, suppl. 2, pp. 223–237 (1982)

Prömel, H.J., Voigt, B.: Canonical partition theorems for parameter sets. J. Comb. Theory A **35**, 309–327 (1983)

Prömel, H.J.: Induced partition properties of combinatorial cubes. J. Combin. Theory. Ser. A **39**, 177–208 (1985)

Prömel, H.J., Voigt, B.: Canonizing Ramsey theorems for finite graphs and hypergraphs. Discrete Math. **54**, 49–59 (1985)

Prömel, H.J., Voigt, B.: Hereditary attributes of surjections and parameter sets. Eur. J. Comb. **7**, 161–170 (1986)

Prömel, H.J., Voigt, B.: A sparse Graham-Rothschild theorem. Trans. Am. Math. Soc. **309**, 113–137 (1988)

Prömel, H.J., Voigt, B.: A short proof of the restricted Ramsey theorem for finite set systems. J. Comb. Theory A **52**, 313–320 (1989)

Prömel, H.J., Voigt, B.: A partition theorem for [0, 1]. Proc. Am. Math. Soc. **109**, 281–285 (1990)

Prömel, H.J., Thumser, W., Voigt, B.: Fast growing functions based on Ramsey theorems. Discrete Math. **95**, 341–358 (1991)

Pudlák, P.: Boolean complexity and Ramsey theorems. In: Mathematics of Ramsey Theory, pp. 246–252. Springer, Berlin (1990)

Pudlák, P., Tůma, J.: Every finite lattice can be embedded in a finite partition lattice. Algebr. Univers. **10**, 74–95 (1980)

Rado, R.: Studien zur Kombinatorik. Math. Z. **36**, 424–470 (1933a)

Rado, R.: Verallgemeinerung eines Satzes von van der Waerden mit Anwendungen auf ein Problem der Zahlentheorie. Sitzungsber. Preuß. Akad. Wiss. Phys.-Math. Klasse **17**, 1–10 (1933b)

Rado, R.: Note on combinatorial analysis. Proc. Lond. Math. Soc. **48**, 122–160 (1943)

Rado, R.: Direct decomposition of partitions. J. Lond. Math. Soc. **29**, 71–83 (1954)

Rado, R.: Universal graphs and universal functions. Acta Arith. **9**, 331–340 (1964)

Rado, R.: Note on canonical partitions. Bull. Lond. Math. Soc. **18**, 123–126 (1986)

Radziszowski, S.: Small Ramsey numbers. Electron. J. Comb. Dyn. Surv. **DS1** (2011)

Ramsey, F.P.: Critical note of L. Wittgenstein's Tractatus Logico-Philosophicus. Mind. N.S. **32**, 465–478 (1923)

Ramsey, F.P.: The foundations of mathematics. Proc. Lond. Math. Soc. **S2–25**, 338–384 (1926a)

Ramsey, F.P.: Mathematical logic. Math. Gaz. **13**, 185–194 (1926b)

Ramsey, F.P.: A contribution to the theory of taxation. Econ. J. **37**, 47–61 (1927)

Ramsey, F.P.: A mathematical theory of saving. Econ. J. **38**, 543–549 (1928)

Ramsey, F.P.: On a problem of formal logic. Proc. Lond. Math. Soc. **S2–30**, 264–286 (1930)

Ramsey, F.P.: In: Mellor, D.H. (ed.) Foundations. Routledge & Kegan Paul, London (1978)

Rödl, V.: On Ramsey families of sets. Graphs Comb. **6**, 187–195 (1990)

Rödl, V., Ruciński, A.: Lower bounds on probability thresholds for Ramsey properties. In: Combinatorics, Paul Erdős Is Eighty. Bolyai Society Mathematics Studies, vol. 1, pp. 317–346. János Bolyai Mathematical Society, Budapest (1993)

Rödl, V., Ruciński, A.: Random graphs with monochromatic triangles in every edge coloring. Random Struct. Algorithms **5**, 253–270 (1994)

Rödl, V., Ruciński, A.: Threshold functions for Ramsey properties. J. Am. Math. Soc. **8**, 917–942 (1995)

Rödl, V., Skokan, J.: Regularity lemma for k-uniform hypergraphs. Random Struct. Algorithms **25**, 1–42 (2004)

Roth, K.F.: On certain sets of integers. J. Lond. Math. Soc. **28**, 104–109 (1953)

Sanders, J.H.: A Generalization of Schur's Theorem. Phd. Thesis, Yale University (1968)

Saxton, D., Thomason, A.: Hypergraph containers (2012). Preprint, arXiv:1204.6595

Schmerl, J.: Problem session, problem 3.5. In: Graphs and Order, pp. 539–540. D. Reidel, Dordrecht (1971)

Schmerl, J.H.: Problem session, Problem 3.5. In: Rival, I. (ed.) Graphs and Order, pp. 539–540. D. Reidel, Dordrecht (1985)

Schmerl, J.H.: Finite substructure lattices of models of Peano arithmetic. Proc. Am. Math. Soc. **117**, 833–838 (1993)

Schur, I.: Über die Kongruenz $x^m + y^m = z^m \pmod{p}$. Jahresber. Dtsch. Math.-Verein. **25**, 114–117 (1916)

Schütte, K.: Proof Theory. Springer, Berlin (1977). Translated from the revised German edition by J. N. Crossley

Shearer, J.B.: A note on the independence number of triangle-free graphs. Discrete Math. **46**, 83–87 (1983)

Shelah, S.: Primitive recursive bounds for van der Waerden numbers. J. Am. Math. Soc. **1**, 683–697 (1988)

Sierpiński, W.: Sur un problème de la théorie des relations. Ann. Sc. Norm. Sup. Pisa Cl. Sci. **2**, 285–287 (1933)

Sierpiński, W., Tarski, A.: Sur une propriété caracteristique des nombres inaccessibles. Fund. Math. **15**, 292–300 (1930)

Skolem, T.: Ein kombinatorischer Satz mit Anwendung auf ein logisches Entscheidungsproblem. Fund. Math. **20**, 254–261 (1933)

Sós, V.T.: Irregularities of partitions: Ramsey theory, uniform distribution. In: Surveys in Combinatorics (Southampton, 1983), pp. 201–246. Cambridge University Press, Cambridge (1983)

Spencer, J.: Ramsey's theorem – a new lower bound. J. Comb. Theory A **18**, 108–115 (1975a)

Spencer, J.: Restricted Ramsey configurations. J. Comb. Theory A **19**, 278–286 (1975b)

Spencer, J.: Asymptotic lower bounds for Ramsey functions. Discrete Math. **20**, 69–76 (1977)

Spencer, J.: Canonical configurations. J. Comb. Theory A **34**, 325–330 (1983)

Spencer, J.: Probabilistic methods. Graphs Comb. **1**, 357–382 (1985)

Spencer, J.: Counting extensions. J. Comb. Theory A **55**, 247–255 (1990)

Sperner, E.: Ein Satz über Untermengen einer endlichen Menge. Mathematische Zeitschrift **27**, 544–548 (1928)

Stevens, R.S., Shantaram, R.: Computer-generated van der Waerden partitions. Math. Comp. **32**, 635–636 (1978)

Szekeres, G.: A combinatorial problem in geometry – reminescences. In: Paul Erdős, The Art of Counting, pp. xix–xxii. MIT, Cambridge (1973)

Szemerédi, E.: On sets of integers containing no four elements in arithmetic progression. Acta Math. Acad. Sci. Hung. **20**, 89–104 (1969)

Szemerédi, E.: On sets of integers containing no k elements in arithmetic progression. Acta Arith. **27**, 199–245 (1975)

Taylor, A.D.: A canonical partition relation for finite subsets of ω. J. Comb. Theory A **21**, 137–146 (1976)

Taylor, A.D.: Bounds for the disjoint unions theorem. J. Comb. Theory A **30**, 339–344 (1981)

Taylor, A.D.: A note on van der Waerden's theorem. J. Comb. Theory A **33**, 215–219 (1982)

van der Waerden, B.L.: Beweis einer Baudetschen Vermutung. Nieuw Arch. Wisk. **15**, 212–216 (1927)

van der Waerden, B.L.: Einfall und Überlegung in der Mathematik – Der Beweis der Vermutung von Baudet. Elem. Math. **9**, 49–56 (1954)

van der Waerden, B.L.: How the proof of Baudet's conjecture was found. In: Studies in Pure Mathematics (Presented to Richard Rado), pp. 251–260. Academic, London (1971)

Voigt, B.: The partition problem for finite Abelian groups. J. Comb. Theory A **28**, 257–271 (1980)

Voigt, B.: Canonizing partition theorems: diversification, products, and iterated versions. J. Comb. Theory A **40**, 349–376 (1985)

Wainer, S.S.: A classification of the ordinal recursive functions. Arch. Math. Log. Grundl. **13**, 136–153 (1970)

Wainer, S.S.: Ordinal recursion, and a refinement of the extended Grzegorczyk hierarchy. J. Symb. Log. **37**, 281–292 (1972)

Witt, E.: Ein kombinatorischer Satz der Elementargeometrie. Math. Nachr. **6**, 261–262 (1952)

Zykov, A.A.: On some properties of linear complexes. Am. Math. Soc. Transl. **1952**, 33 (1952)

Index

H.J. Prömel, *Ramsey Theory for Discrete Structures*,
DOI 10.1007/978-3-319-01315-2,
© Springer International Publishing Switzerland 2013

Printed in the United States
By Bookmasters